X-15
PHOTO SCRAPBOOK

COMPILED BY

TONY R. LANDIS AND DENNIS R. JENKINS

specialty press
PUBLISHERS AND WHOLESALERS

© 2003 by Tony R. Landis and Dennis R. Jenkins

All rights reserved. No part of this publication may be reproduced or utilized in any form or by any means, electronic or mechanical, including photocopying, recording, or by any information storage and retrieval system, without prior written permission from the author. All photos and artwork are the property of the owners as credited.

The information in this work is true and complete to the best of our knowledge. However, all information is presented without any guarantee on the part of the authors or publisher, who also disclaim any liability incurred in connection with the use of the information.

In Memory of

Raymond and Donna Landis

ISBN 1-58007-074-4

Item Number SP074

39966 Grand Avenue
North Branch, MN 55056 USA
(651) 277-1400 or (800) 895-4585
www.specialtypress.com

Printed in China

Distributed in the UK and Europe by:

Midland Publishing
4 Watling Drive
Hinckley LE10 3EY, England
Tel: 01455 233 747 Fax: 01455 233 737
www.midlandcountiessuperstore.com

On the Cover: *Flight 2-17-33 on 23 June 1961, piloted by Bob White, was the first Mach 5 excursion by any piloted aircraft. Balls Three is overhead.* (NASA Dryden)

At upper left on the Back Cover: *Bob Rushworth took Flight 2-28-48 on 29 August 1962 to Mach 5.12 and 97,200 feet. The single XLR99 rocket engine burned for 92 seconds before it ran out of propellants.* (NASA Dryden)

At upper right on the Back Cover: *When it was carrying its large external tanks, the X-15A-2 was the most colorful of the three airplanes. This is Flight 2-50-89 on 18 November 1966.* (NASA Dryden)

At left on the Back Cover: *Scott Crossfield poses with the X-15-1 prior to the first captive flight.* (The Boeing Company Archives)

On the Title Page: *The X-15-1 in front of the North American Aviation facility in Inglewood, California. The X-15s were small enough to be transported over the open roads; the only disassembly required was removing the dorsal rudder. Note the large neon sign on the North American building proclaiming its involvement with the X-15 program, a source of pride to those involved.* (Boeing via the Gerald H. Balzer Collection)

INTRODUCTION

The North American X-15 was the last in a line of manned rocket-powered research airplanes built during the 1950s to explore ever-faster and higher flight regimes. This was an era before computers were commonplace, and the only way to investigate the unknown was to go there. The program was begun in 1954 specifically to produce the first hypersonic (velocities greater than five times the speed of sound) manned aircraft. Forward-thinking researchers also designed the airplane to fly to the edge of space, long before the manned space program had begun in earnest. The National Advisory Committee on Aeronautics (NACA) developed a concept, and the Air Force and Navy eventually funded North American Aviation to build three flight vehicles.

The X-15 proved to be the most successful of the high-speed research airplanes that were tested at Edwards Air Force Base, California, between 1946 and 1968. The three airplanes completed 199 flights and exceeded all of the design goals. The program recorded the fastest manned aircraft flight when Major William J. "Pete" Knight took the modified X-15A-2 to 4,520 mph (Mach 6.7). Similarly, Joseph A. Walker set an unofficial altitude record for manned aircraft of 354,200 feet (67 miles). The only piloted winged vehicles to exceed either of these records in the 40 years since are the United States' five Space Shuttle Orbiters.

In a bizarre quirk of record-keeping, another altitude milestone, set by Major Robert M. White, still stands. The Fédération Aéronautique Internationale (FAI), which is the governing body for aviation records, defines the atmosphere ending at 328,099 feet (100 kilometers). White climbed to 314,750 feet. According to the rules, new records must exceed the old by 3 percent, meaning somebody will have to fly at least 324,193 feet altitude to beat it. But the attempt can not exceed 328,099 feet (as with Joe Walker's 354,200-foot flight), else it will be disqualified since it is outside the atmosphere. It has been over four decades and nobody has tried yet.

An in-depth history of the X-15 program may be found in *Hypersonic: The Story of the North American X-15* by Dennis R. Jenkins and Tony R. Landis (ISBN 1-58007-68-X, Specialty Press, 39966 Grand Avenue, North Branch, MN, 55056, (651) 277-1400 or (800) 895-4585, or visit http://www.specialtypress.com). This scrapbook began as a collection of illustrations that were assembled for *Hypersonic* but would not fit into the finished work. Since many of these are significant – or at least interesting – and most have never been published previously, it was decided to print the scrapbook as a companion volume to *Hypersonic*. However, as we worked on the scrapbook, additional images were discovered, including several contributed by former *National Geographic* photographer Dean Conger.

As always, we are very grateful to several institutions and individuals who supplied photos for this publication: Michael J. Lombardi at The Boeing Company Archives, the AFFTC History Office, the Dryden Flight Research Center, Anthony Accurso, Gerald H. Balzer, Dean Conger, Benjamin F. Guenther, Jay Miller, and Terry Panopalis.

Most of the photos presented here are of high quality, but a few are less than ideal; they are included because they offer a glimpse of the program that the authors believed was worthy of being seen. The captions run the gamut from long and detailed, to almost trivial; sometimes the photo was not well annotated and we have little idea of what was being done by whom, but still felt the photo needed to be published. The goal was to present an interesting visual record of the program. We hope you enjoy it.

Tony R. Landis
Lancaster, California

Dennis R. Jenkins
Cape Canaveral, Florida

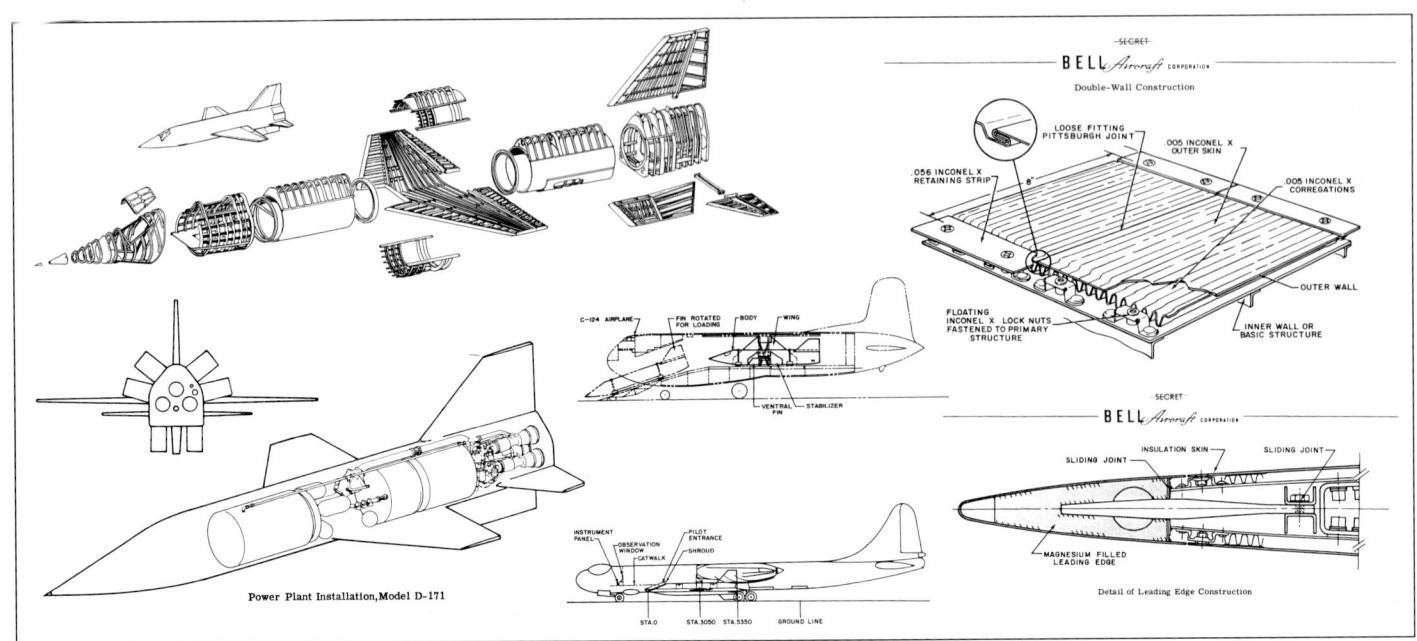

The Bell D171 entry in the Project 1226 competition. Note the use of a Douglas C-124 to transport the aircraft to Edwards, and a Convair B-36 as a carrier aircraft. Bell used "loose fitting Pittsburgh joints" in most of the outer skin. (Courtesy of Benjamin F. Guenther)

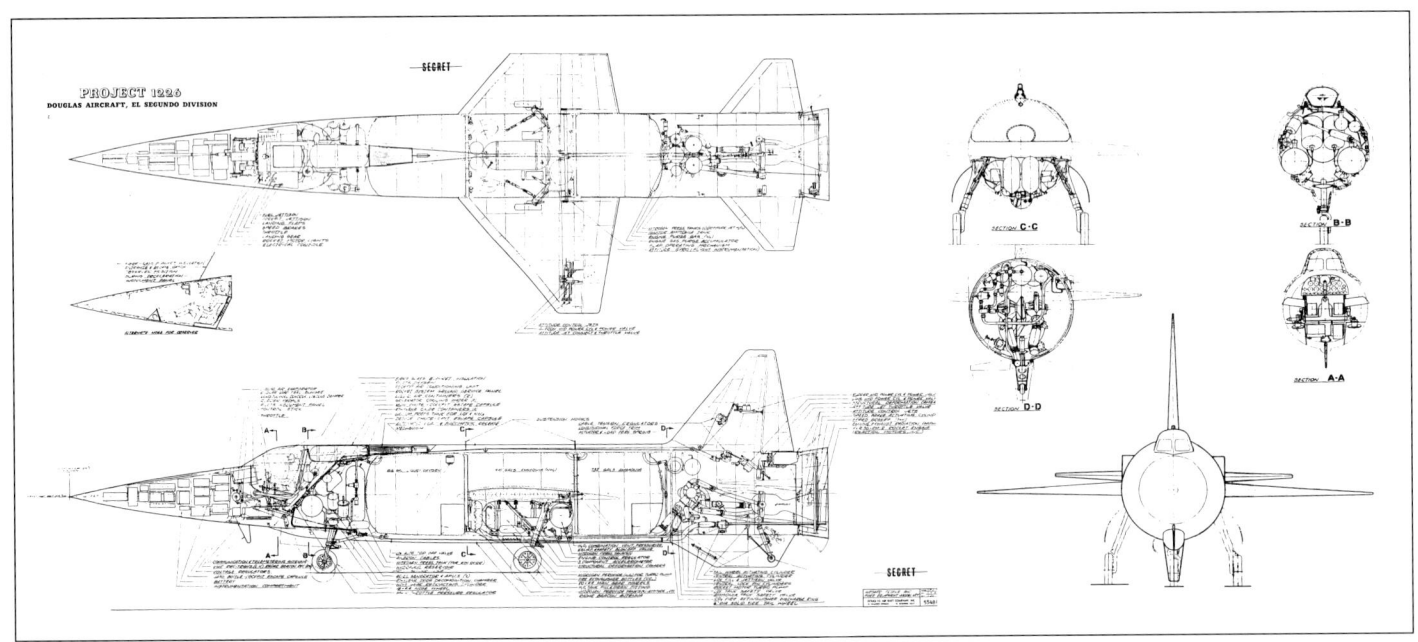

Douglas wrote a good proposal, but included remarkably little artwork. The design had evolved considerably from the Model 671 ("D-558-III") developed earlier for the Navy. Douglas placed second, just slightly behind North American. (Courtesy of Benjamin F. Guenther)

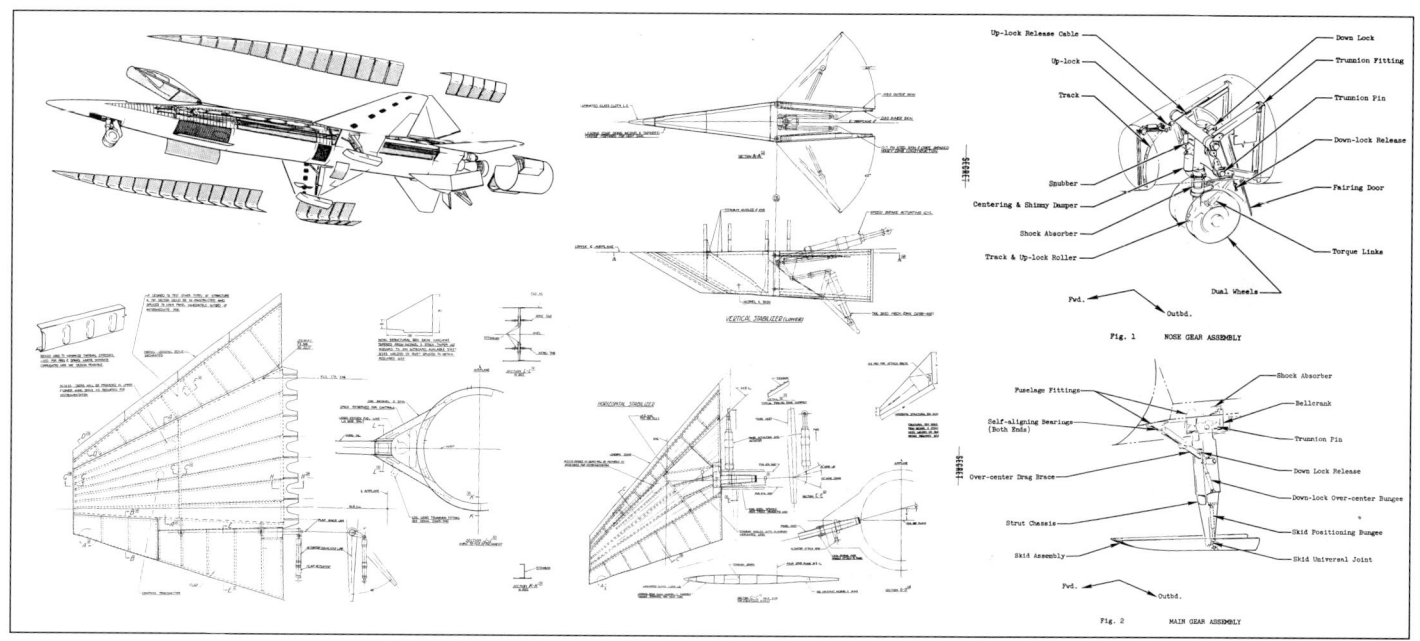

The North American design carried an awkward ESO-7487 moniker, but was the closest in concept to what the NACA researchers at Langley had envisioned. Note the fuselage tunnels that extend all the way to the nose (top left). (Courtesy of Benjamin F. Guenther)

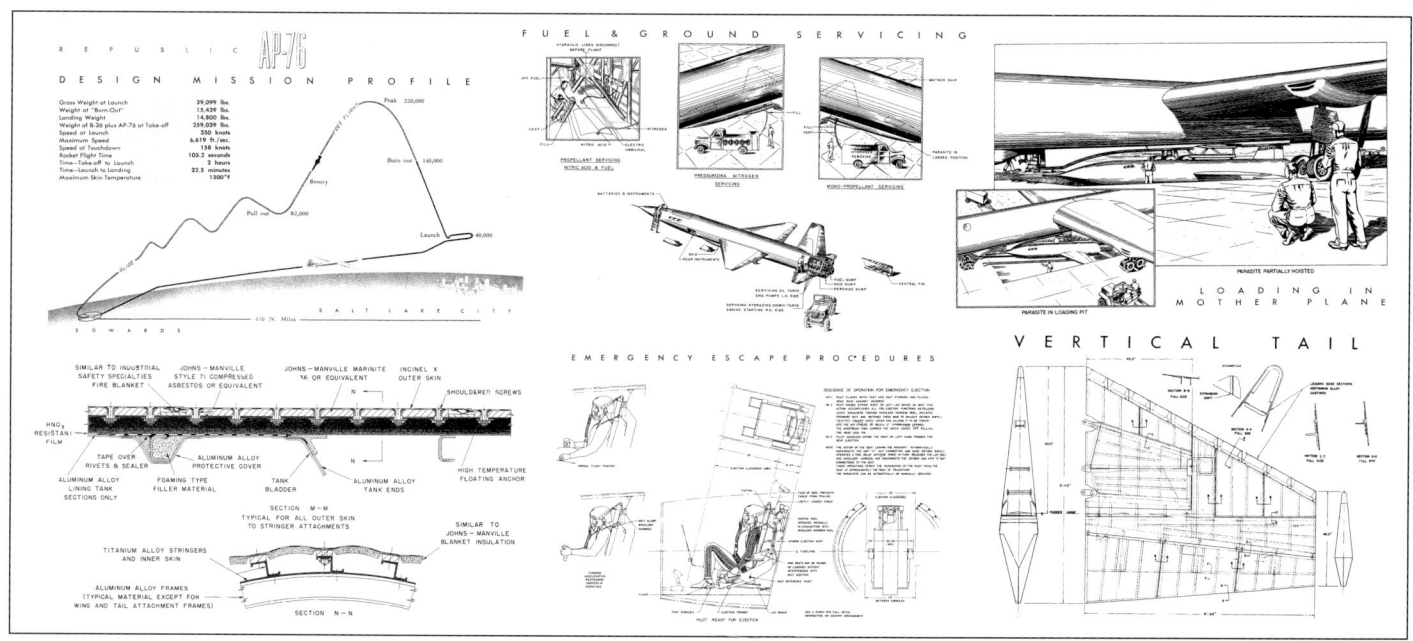

The Republic AP-76 was the heavyweight of the competitors, and included several lessons-learned from the still-born XF-103 interceptor program, particularly involving the cockpit and pilot accommodations. This design placed last in the evaluation. (Courtesy of Benjamin F. Guenther)

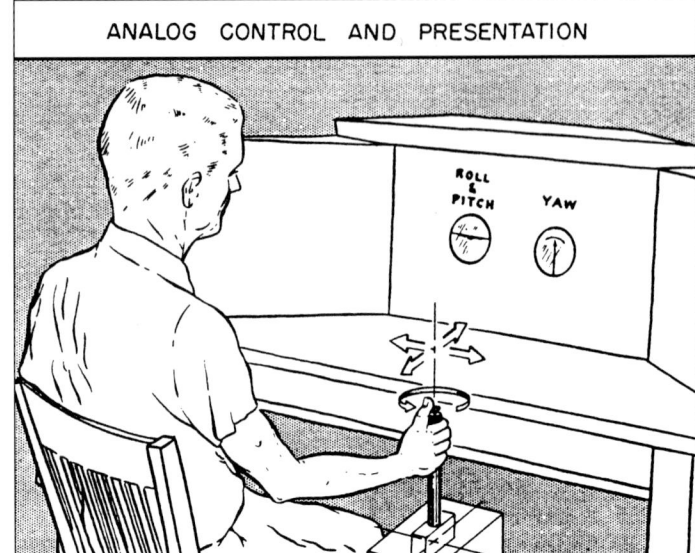

During the mid-1950s, aircraft simulators were far from the advanced devices in use today. At left, Richard E. Day sits in front of the General Electric Differential Analyzer (analog computer) fitted with an airplane stick and controls to create a crude simulator for the Bell X-2. The illustration above is an early (1956) simulator used during the X-15 development. (NASA Dryden)

The Aviation Medical Acceleration Laboratory (AMAL) at the Naval Air Development Center (NADC) Johnsville, Pennsylvania, played a major role in training the original cadre of X-15 pilots using a large human centrifuge. The basic techniques established by the X-15 program were later used to train the Mercury Seven at the same facility. (U.S. Navy)

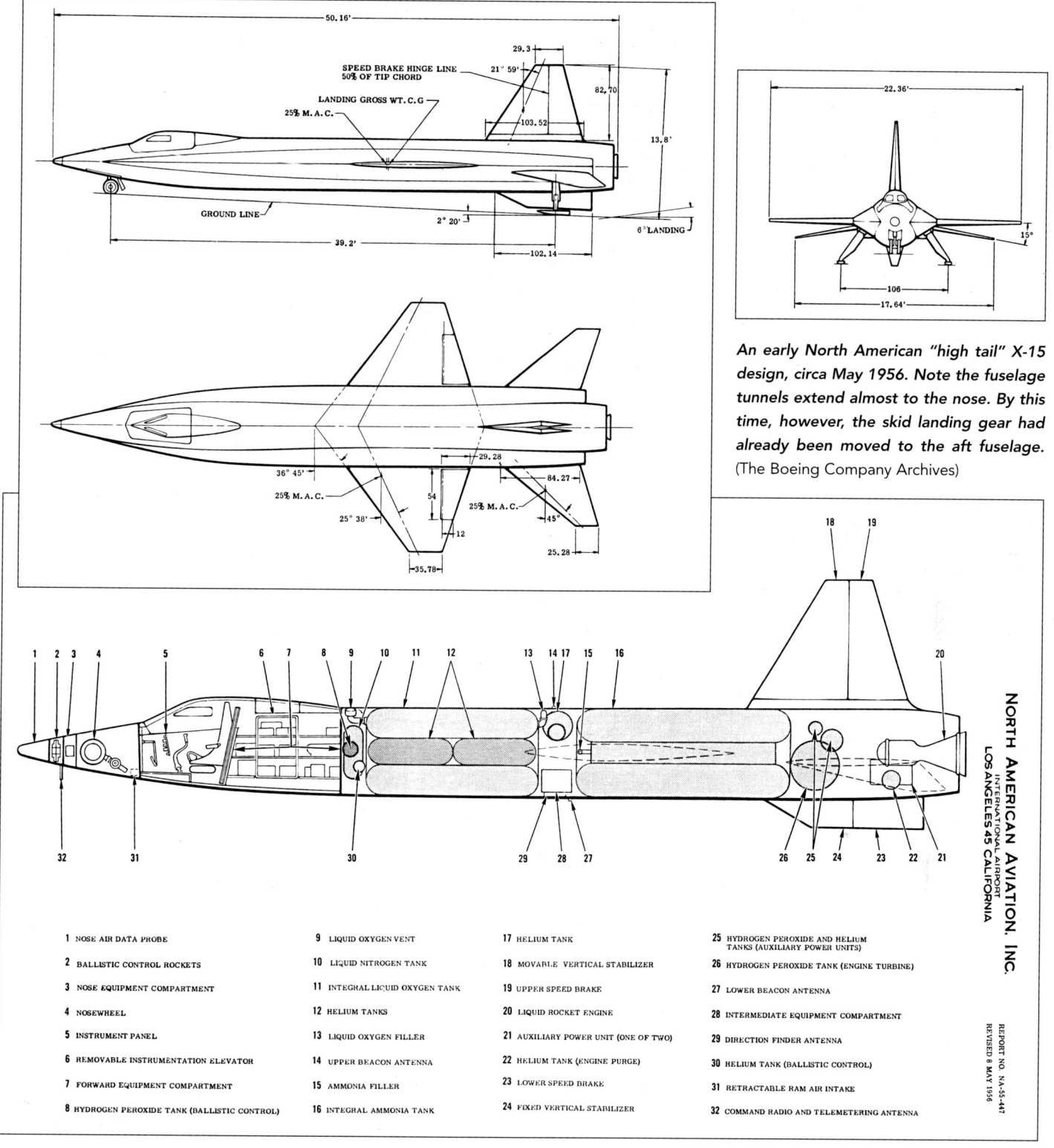

An early North American "high tail" X-15 design, circa May 1956. Note the fuselage tunnels extend almost to the nose. By this time, however, the skid landing gear had already been moved to the aft fuselage. (The Boeing Company Archives)

X-15 Photo Scrapbook

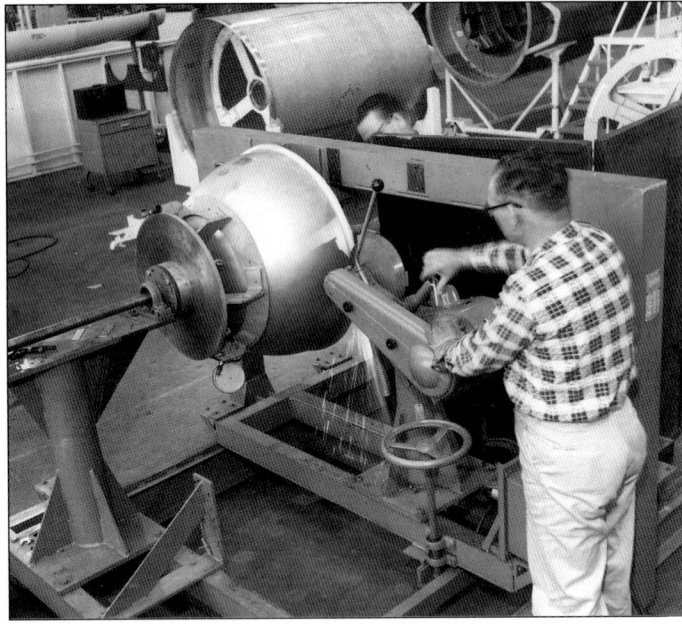

Not surprisingly, a great deal of the construction process involved making the various propellant tanks. In addition to the main liquid oxygen and ammonia tanks, containers were needed for hydrogen peroxide, helium, and nitrogen. Here a worker works on one of the smaller gas tanks. (Boeing via the Gerald H. Balzer Collection)

Models of the Project 1226 competitors give a good indication of their relative sizes. The Bell X-1A is in the middle as a reference. Clockwise from lower left: Bell D171, North American ESO-7487, Republic AP-76, and Douglas Model 684. (Terry Panopalis Collection)

North American decided that using full-monocoque propellant tanks would save a little weight and greatly ease the thermal problems associated with the cold propellants and heat transfer from high-speed flight. This is one of the main tanks in its manufacturing jig. Note the man's head protruding from the middle of the tank. (Boeing via the Gerald H. Balzer Collection)

The extreme aft fuselage of X-15-1 on 30 January 1958. Note that the fuselage tunnels end flush with the structural fuselage; a non-structural fairing would extend further aft to cover the engine nozzle(s) after they were installed. (Boeing via the Gerald H. Balzer Collection)

The X-15-1 during final assembly. Note the calibration device around the upper speed brakes. The person inside the engine compartment gives a good indication of the size of the fuselage. (Boeing via the Gerald H. Balzer Collection)

The wing for X-15-1 in the assembly jig shows the corrugated rib shear webs, a design feature that allowed the structure to accommodate thermal expansion. This type of construction is known as a hot-structure. (Boeing via the Gerald H. Balzer Collection)

In an era before CAD/CAM took over, drawings were prepared by draftsmen using pencils and ink in rooms filled with drafting tables. Note the "X-15" sign hanging from the ceiling and Scott Crossfield talking to the draftsman. (Boeing via the Gerald H. Balzer Collection)

The X-15 ejection seat and David Clark MC-2 full-pressure suit were tested on this special "wind blast" rig at the South base sled track during early 1959. Other wind blast tests were conducted in the wind tunnels at NACA Ames. (AFFTC History Office via the Gerald H. Balzer Collection)

The South Base sled track was also used for separation and deployment tests of the ejection seat itself. A cockpit section would be accelerated to high speed and high dynamic pressure by a set of rockets, then the ejection sequence would be started (note the canopy separating in below, center). Initial tests were less than successful, as often as not because of problems with the sled or its rockets. Eventually the testing would be cut short, but sufficient data had be acquired to clear the seat for use in the X-15. (The Boeing Company Archives)

10   X-15 Photo Scrapbook

The final seat run of the wind blast sled, conducted on 14 March 1958. The number of rocket bottles in the sled could be varied to tailor the dynamic conditions for each test. (AFFTC History Office)

After the seat was qualified, another run was made on 17 March 1958 to qualify the MC-2 full-pressure suit. This run was apparently made to much higher speeds and dynamic pressures, at least judging by the extra rocket motors on the second carrier. (AFFTC History Office)

Two shots of a successful X-15 ejection system test that included the seat and the David Clark full-pressure suit. The system was qualified for use at speeds up to Mach 4 and altitudes up to 120,000 feet. (AFFTC History Office)

An instrumented anthropomorphic dummy was used during the ejection tests. This photo shows the special 24-foot diameter parachute developed for the X-15 program; later, standard Air Force 28-foot chutes would be used as replacements. (AFFTC History Office)

This and Facing Page: *Scenes from the X-15 roll-out ceremony, including Vice President Richard M. Nixon with North American president James H. "Dutch" Kindelberger (at right). Sometime before the ceremony Scott Crossfield posed for the photo below. Note the upper bug-eye camera fairings just behind the canopy.* (The Boeing Company Archives)

66670
U.S. AIR FORCE
FUEL VENT
APU EXHAUST
HYDROGEN PEROXIDE
JETTISON VENT
X-15

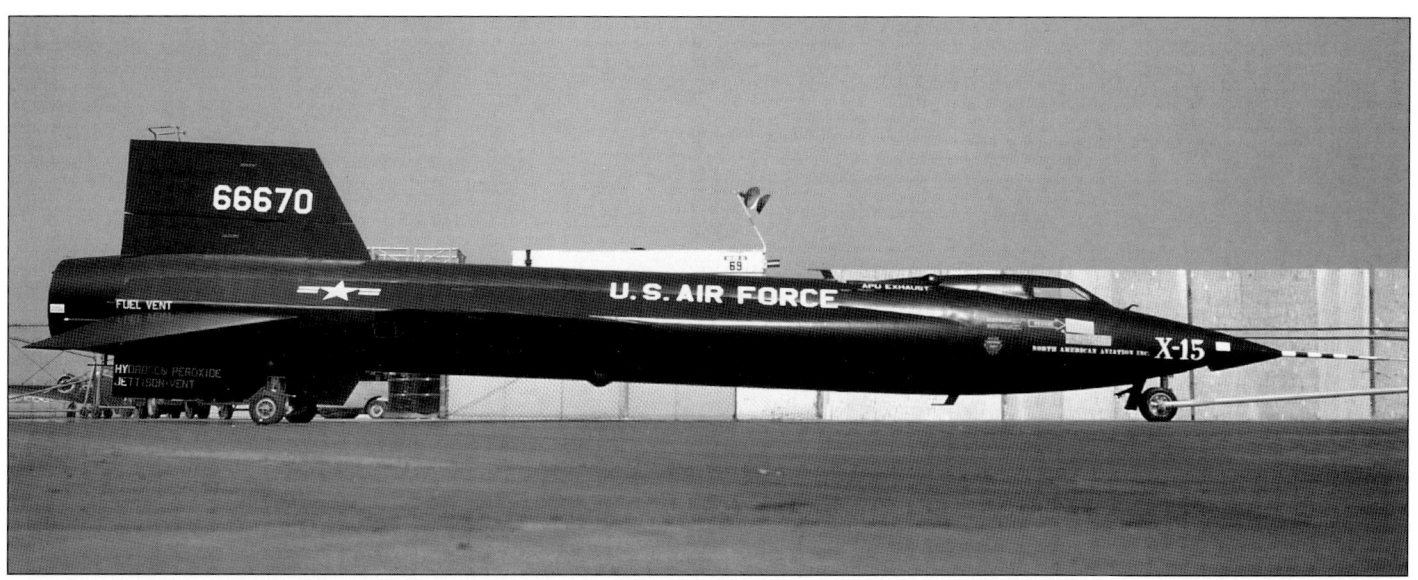

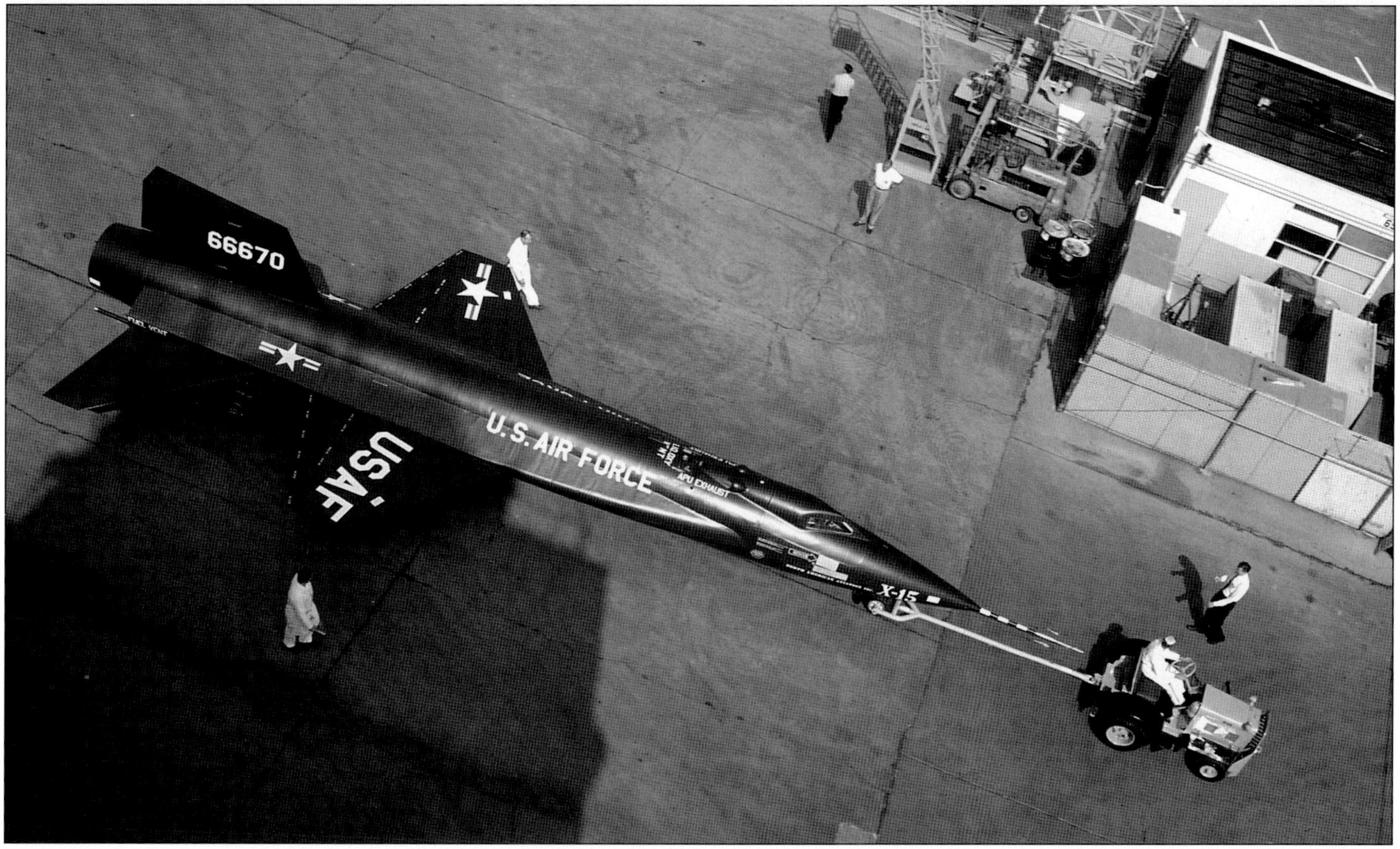
66670
U.S. AIR FORCE
USAF
APU EXHAUST
X-15

The first three men to fly the X-15 were, from left: Joseph A. Walker from NASA, Captain Robert M. White from the Air Force, and Scott Crossfield from North American Aviation. This photo was taken just prior to the roll-out ceremony. (The Boeing Company Archives)

Sometime prior to roll-out in October 1958, Alvin S. White showed the X-15-1 to King Baudouin of Belgium. (Courtesy of Alvin S. White)

North American kept White busy, shown here with General Napoli, Chief of Staff of the Italian Air Force. (Courtesy of Alvin S. White)

The fixed-base simulator in Inglewood. Most early flights would be planned and practiced using the North American simulator since the one at the Flight Research Center did not yet have a full set of analog computers. (The Boeing Company Archives)

One of the more innovative aspects of the X-15 was its side-stick controllers. From left, Al White, Iven Kincheloe, and Scott Crossfield discuss the merits of one proposed design for the controller hand grip. (Courtesy of Alvin S. White)

Both Al White and Bob White had flown with the 354th Fighter Group, and are shown here with a model of a P-51 Mustang. The two men were unrelated, except by name, and the fact that both were selected as X-15 pilots. (Courtesy of Alvin S. White)

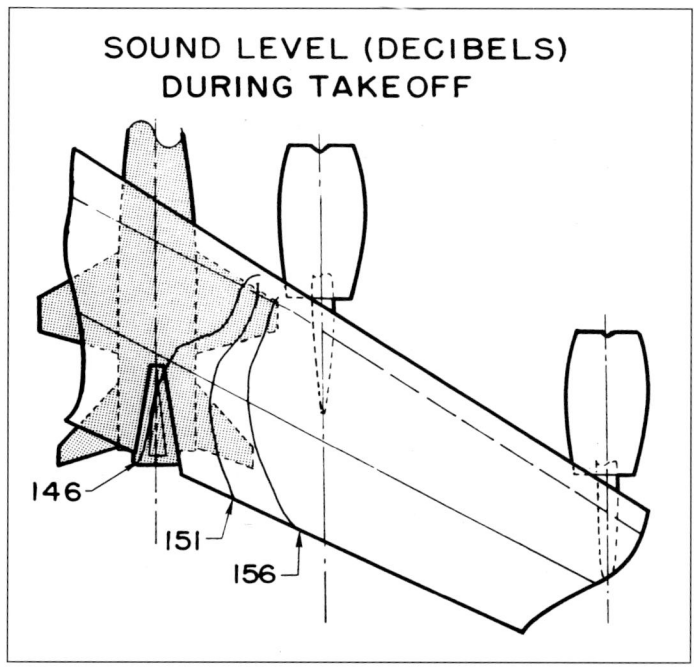

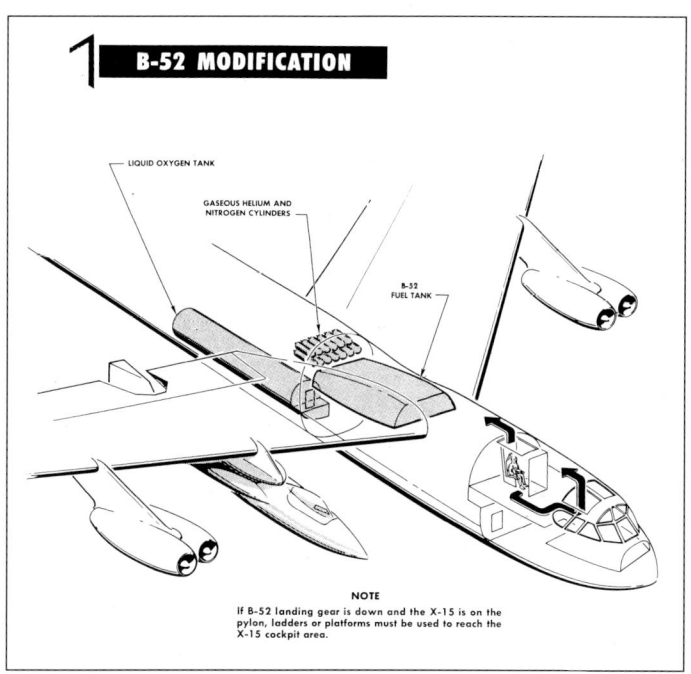

A major concern early in the program was the effect of noise from the NB-52 Nos. 5 and 6 engines on the X-15 structure. Tests and early flights revealed that several portions of the X-15 empennage needed to be strengthened to survive the environment. (U.S. Air Force)

A drawing showing some of the major modifications to the two NB-52s. There were many additional minor changes to the bombers. The launch panel operator had two escape routes, one through an upper hatch and the other through the flight deck. (U.S. Air Force)

Neil Armstrong, Iven Kincheloe, and Bob White at the South Base sled track in 1959 during X-15 ejection seat qualification. Kinch would be killed before he got a chance to fly the X-15. (AFFTC History Office)

Flight testing the NB-52A was done by Captain John E. "Jack" Allavie (left) and Captain Charles C. "Charlie" Bock (right). The person in the middle was not identified. (AFFTC History Office)

X-15 Photo Scrapbook

The X-15 was a small airplane, and very little disassembly was required for the overland trip from Inglewood to Edwards; mainly removing the ball nose and the dorsal and ventral rudders. Here is X-15-1 being unwrapped following its delivery on 17 October 1958. (AFFTC History Office)

The mockup in Inglewood in its final form. Note the canopy sitting on the ground beside the fuselage, and the access stand over the horizontal stabilizer to allow the engine compartment to be examined. (The Boeing Company Archives)

Al White would never get to fly the X-15, but was frequently used as a stand-in for Scott Crossfield during public appearances. Here he is showing the X-15 mockup at a Boy Scout Jamboree in downtown Los Angeles. (Courtesy of Alvin S. White)

A publicity photo of Crossfield, Walker, and Kincheloe looking at the X-15 mockup in Inglewood. Note the shape of the upper bug-eye camera fairings just behind the canopy, and the spotlights on the inside of the canopy. (The Boeing Company Archives)

Workers prepare a 1/10-scale model of the X-15 for testing in the 7 x 10-foot wind tunnel at NASA Langley. This series of tests was related to the spin characteristics of the design. (NASA Langley)

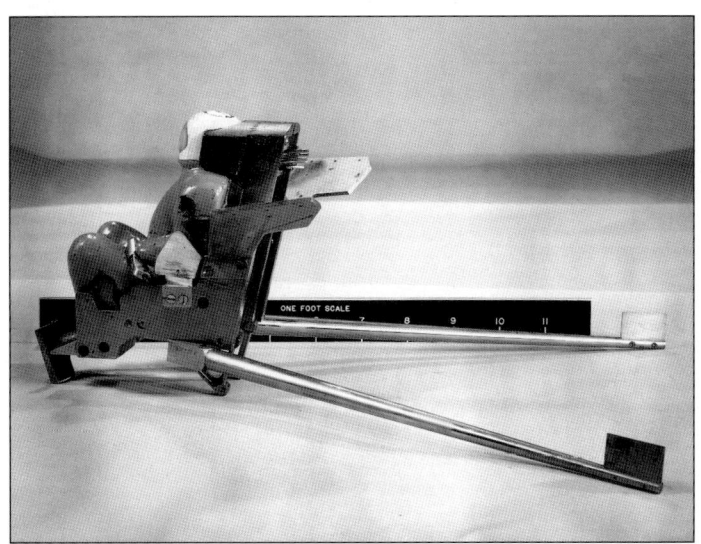

Everything had a wind tunnel model built. This is a small model of the X-15 ejection seat showing the stabilization booms and "wings" in their deployed positions. The photo was dated 20 August 1958. (The Boeing Company Archives)

Reaction controls were a new concept in the late 1950s. To gain some experience with the control system, a JF-104A (55-2961) was equipped with thrusters in special wing-tip fairings and also in the nose. The airplane is shown here on 4 August 1959. (NASA Dryden)

A series of heating tests were conducted on samples of the X-15 fuselage to evaluate the effects of temperatures on the integral propellant tanks, and also on the Inconel X side panel fairings. Other tests were run on the wing and stabilizers. Here a group that includes Scott Crossfield examines the test fixture while it is inactive (at left, above), and again while it is heating (note the dark visors they are holding). The final photo (right) is at the conclusion of the test. (The Boeing Company Archives)

RAF Squadron Leader Harry M. Archer (left) poses with Jack Allavie. Archer flew in the NB-52 during several missions while he was on an exchange program. (AFFTC History Office via the Terry Panopalis Collection)

Engineers and technicians replace the XLR99 engine in the X-15-3 sometime during 1961. Note the man inside the engine compartment in the photo at right. (Reaction Motors via the Terry Panopalis Collection)

18 X-15 Photo Scrapbook

The Rocket Engine Test Facility at Edwards allowed engine runs at full thrust on the ground. This is X-15-1 in the most often used engine run area on 5 February 1959. It was not particularly unusual for an X-15 to be in this area, while the Propulsion System Test Stand (a set of propellant tanks and thrust structure) was in the other test area certifying an XLR99 for flight. Occasionally, an X-15 would be located in each test area if two flights were scheduled close together. The concrete control bunker is at the top center. (AFFTC History Office)

The X-15-1 being mated to the NB-52A. Note the open entrance hatch on the carrier aircraft. The mating operation was not terribly complex and only took a couple of hours. (AFFTC History Office via the Gerald H. Balzer Collection)

North American, Boeing, and the Air Force conducted a series of vibration tests after the NB-52A was modified to ensure the natural frequency of the pylon did not damage either the B-52 or the X-15. (Boeing via the Gerald H. Balzer Collection)

Scott Crossfield climbs out of the X-15-1 after the first captive carry flight. The X-15 landing gear had been deployed during the flight to demonstrate it would work after being cold-soaked at altitude. (AFFTC History Office via the Gerald H. Balzer Collection)

An unusual view of the X-15-1 being mated to Balls Three. Note the "remove before flight" streamer in the nose gear well, and that the lower BCS thruster is offset to one side (the top one was offset to the other side). (AFFTC History Office via the Gerald H. Balzer Collection)

Two views of the X-15-1 at the Rocket Engine Test Facility. Hydrogen peroxide to power the turbopumps is being loaded into the airplane at right; note the lack of special protective suits on the ground crew. Peroxide was not considered particularly dangerous, but still needed to be handled carefully. (AFFTC History Office via the Gerald H. Balzer Collection)

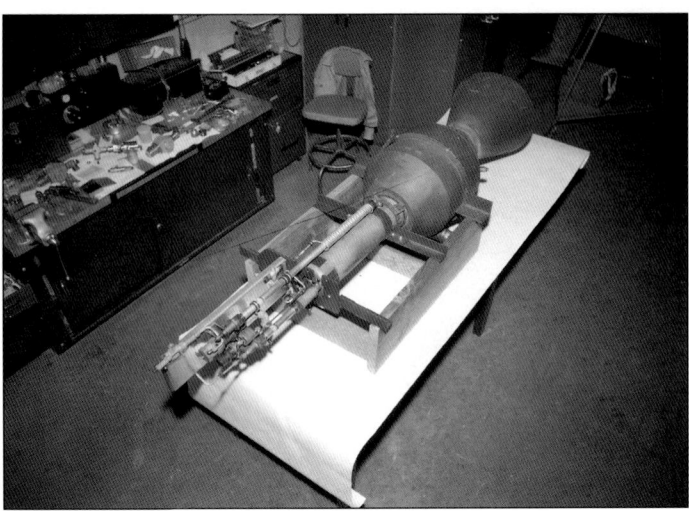

With seven of eight chambers lit, the X-15-2 accelerates away during Flight 2-5-12 on 17 February 1960 with Scott Crossfield in the cockpit. Initially, the upper engine would not light at all, although three chambers ignited during the restart. The last chamber refused to light, so Crossfield kept the other seven burning for 309.2 seconds instead of the 260 seconds that had been planned if all chambers had worked. Despite the extra time, the flight only managed Mach 1.57 instead of the intended Mach 2.0, although Crossfield overshot his expected altitude slightly, reaching 52,640 feet. (AFFTC History Office)

Prior to the arrival of an XLR99 at Edwards, the engineers at the Flight Research Center were concerned that fuel would pool in the igniter (forward) section of the nozzle and possibly explode upon ignition. To test this theory, NASA engineer Wes Messing designed a boilerplate nozzle that was constructed with the help of Edward Saltzman. They attached it to an XLR8/11 chamber and fired it several times at the rocket engine test site with various amounts of fuel placed in the forward section. The fuel never exploded. The photo is dated 10 April 1957. (NASA Dryden)

A photo from an early loading operation. Note another photographer in the background past the step ladder. Hydraulic lifts built into the ramp near the Rocket Engine Test Facility were initially used to raise the X-15 into position under the NB-52. (Jay Miller Collection)

Another early loading operation, but by now a set of stripes had been applied to the (probably X-15-2) forward fuselage to denote the position of the nose landing gear well. The work stands were standard Air Force issue equipment. (AFFTC History Office via the Gerald H. Balzer Collection)

All of the landing gear had to be retracted by hand after the airplane was attached to the NB-52. The nose gear was rather innovative, and was stored with the strut in a compressed state in order to save space in the fuselage. (San Diego Aerospace Museum Collection)

Very few photos were taken from this angle; most were taken from the other side where the X-15 was more visible. This is the first captive flight on 10 March 1959. The X-15 landing gear is down, meaning this was taken in the landing pattern. (AFFTC History Office)

The F-100A used by Scott Crossfield and Al White as a landing trainer. The drag created by the extended speed brake and landing gear, plus the drag chute, created a reasonable simulation of the X-15. (U.S. Air Force via the Terry Panopalis Collection)

A group of engineers gather to discuss the first flight. Note Scott Crossfield in his flight suit on the right side of the table, and the ashtrays and cigarettes on the table. (Gerald H. Balzer Collection)

An unidentified gathering of what appear to be newsmen talking to Scott Crossfield. Note the rather bored Air Force captain in the background. (Gerald H. Balzer Collection)

The lack of exhaust nozzles reveals the two engines in the back of X-15-1 are dummy units. This is most likely an early fit-check on the NB-52A. Note that the fairing around the XLR11s extended further aft than it would for the XLR99. (AFFTC History Office)

Bob White pilots the F-104 while Crossfield lands the X-15-2 following Flight 2-2-6 on 17 October 1959. The chase pilots provided reassurance for the test pilot, and also gained skills to help them on their X-15 flights. (AFFTC History Office)

The JTF-102A (54-1354) was used to test the David Clark full-pressure suits, and to introduce the X-15 pilots to working in the new suits. At left is Bob White on 29 January 1959, with Joe Walker at right on 22 October 1958. (AFFTC History Office)

An X-15 fit check on the NB-52A on 1 December 1958. Note the dummy XLR11s and that the ventral rudder is missing. (NASA Dryden)

Scott Crossfield in an MC-2 full-pressure suit reads over his notes on the way to the airplane. (Boeing via the Gerald H. Balzer Collection)

Forrest Petersen on 3 August 1961 wearing a straw hat on top of his pressure suit helmet. No, we don't know why. (NASA Dryden)

X-15-2 in flight with the two XLR11 engines. The frost around the liquid oxygen tank is visible on the upper fuselage. (AFFTC History Office via the Gerald H. Balzer Collection)

Many early X-15 flights carried "radiation stacks" that used a material that absorbed radiation at the same rate as human tissue. This one has been damaged by something not described. (NASA Dryden)

The plot boards were state-of-the-art when the High Range was built. The position of the X-15 was recorded by different color pens (one for altitude; another for location) on a preprinted map of the operating area. This photo was taken on 17 September 1959. (NASA Dryden)

High Range operations during an abort (1-A-3) on 10 April 1959. This is the position where the NASA-1 communicator (almost always another X-15 pilot) sat in the control room. The plotting boards were located along the walls to the left and behind. (NASA Dryden)

Balls Three and the X-15-2 coming back to Edwards after a weather abort (2-A-8) on 31 October 1959. (AFFTC History Office)

X-15-2 about to touch down on Rogers Dry Lake. The ventral rudder was jettisoned as the airplane turned onto final then the landing gear was deployed. The rudders were normally refurbished for future flights. (AFFTC History Office via the Gerald H. Balzer Collection)

A pair of XLR11 engines in the Propulsion System Test Stand during a ground run on 9 January 1959. Despite sounding like the name of a building, the PSTS was actually a set of propellant tanks and engine mount that closely resembled an X-15. (AFFTC History Office)

An engine explosion during a ground run of an XLR99 on 9 June 1960 essentially destroyed the X-15-3 before its first flight. The photo at right shows the engine compartment after the wreck was removed from the Rocket Engine Test Facility. A decision was made to rebuild the airplane with the advanced Minneapolis-Honeywell MH-96 adaptive flight control system. The addition of the MH-96 made the X-15-3 the airplane of choice for most of the high altitude missions. (AFFTC History Office)

Scott Crossfield looks on as ground crewmen examine the alignment of the X-15 ground handling dolly to the hydraulic lifting jack; barely visible above the jack is the retracted main gear skid. (AFFTC History Office via the Gerald H. Balzer Collection)

The X-15-2 sits on Rogers Dry Lake at the conclusion of the first XLR99-powered flight (2-10-21) on 15 November 1960. Note the Scott Airpack on the suited fireman in the foreground, and the typical 1950s vehicles in the background. (AFFTC History Office)

The X-15-2 before its first powered flight, probably on either 2-C-1 or 2-A-2. Note that Balls Three does not have the astrodome at the launch operator's position yet, and the relative angles-of-attack of the carrier and research aircraft. Outlines of the original small U.S. AIR FORCE markings can barely be seen just aft of the forward camera fairing. (The Boeing Company Archives)

Bob Allen Jr. servicing a nitrogen bottle on 27 July 1960. Note the handwritten North American assembly number on some of the components. (NASA Dryden)

Joe Walker and Bob White inspect the X-15-1 on 25 February 1960. Note that the data block says "X15" (no hyphen). Only the modified second airplane would be called an X-15A. (NASA Dryden)

Above: Joe Walker prepares for Flight 1-7-12 on 12 May 1960. Note the spotlights on the inside of the canopy. (NASA Dryden)

Left: The X-15-1 receives some minor maintenance from Peter Sterk and Edward Nice on 1 June 1960. The aft fuselage cowling has been removed, showing the chambers on the two XLR11 engines. The smaller tube at the 8-o'clock position of each engine is the turbopump exhaust. (NASA Dryden)

Below: The X-15-1 after the first NASA flight (1-3-8) on 25 March 1960. Joe Walker was in the cockpit. (NASA Dryden)

Flight 1-7-12 on 12 May 1960 would be the X-15's first trip beyond Mach 3, and was also the first launch of any X-plane from an uprange lake (all previous launches had been around the Edwards reservation). Joe Walker would take the X-15-1 to 2,111 mph at 77,882 feet. James E. Love, the NASA program manager at Dryden, is looking at the mated pair prior to the flight. (NASA Dryden)

The cockpit of X-15-1 on 23 August 1960. The "throttle" for the XLR11s are eight switches at the front of the console shown above. The instrument panel at right shows that the large stopwatch used to control the engine burn time has not been added yet. Note the stability augmentation system controls at the top of the center pedestal (the rotary knobs selected the amount of gain) and the aerodynamic side-stick controller on the right console. The BCS side-stick controller is shown above. (NASA Dryden)

X-15 Photo Scrapbook

There were several attempts to launch two X-15 flights on the same day; this one was on 4 November 1960. Bob Rushworth successfully flew Flight 1-16-29, but Scott Crossfield had to abort (2-A-20) when the No. 2 APU shutoff valve failed to operate. (NASA Dryden)

This was the end of Flight 1-16-29, a relatively low and slow (48,900 feet and Mach 1.95) familiarization flight for Bob Rushworth. Note the physiological support van (semi trailer) at left and the fire truck. (NASA Dryden)

Note the corrugated sheets used on the back of the fuselage tunnels on X-15-1 and the rods used to support the fuel and oxidizer jettison tubes. This photo was dated 8 June 1960. (NASA Dryden)

An engineer checks the angle of attack on a model of the X-15 before a wind tunnel test. The X-15 was tested at Ames, Langley, and JPL, as well as numerous company and university facilities. (NASA Ames)

Neil Armstrong would fly the X-15 seven times before he went to the Moon. This was his first flight (1-18-31) on 30 November 1960. Note the pitot probe just behind the ball nose on the lower fuselage. There was some concern early in the program that the location just ahead of the windscreen was not ideal, so various other locations were tested; in the end the original position was retained. (NASA Dryden)

Like all airplanes, the X-15 had to be weighed, and Edwards has a dedicated weight and balance hangar. What is interesting about these photos of X-15-3 is that the airplane has a nose boom on it, but it never flew in that configuration. These were taken on 19 May 1960 – three weeks before the XLR99 explosion – and may be the only surviving photos of the airplane in its original configuration. (Gerald H. Balzer Collection)

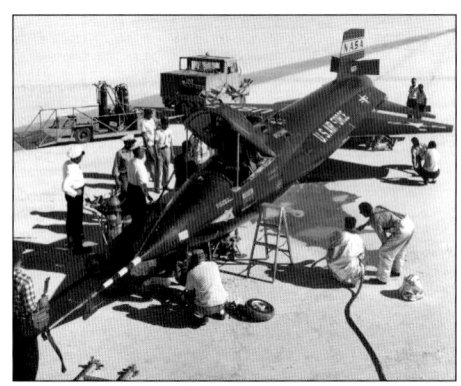  

Three post-flight photos of X-15-1. The shot at right shows Scott Crossfield in an MC-2 full-pressure suit walking to a car to be driven back to the main base; note the water on the lakebed near the aft fuselage, indicating a possible fire after landing. The other two photos show Bob White after an unidentified mission. (AFFTC History Office via the Gerald H. Balzer Collection)

One of the NB-52s with an X-15 flies over a recovery crane on a lakebed. (AFFTC History Office via the Gerald H. Balzer Collection)

Scott Crossfield, Bob White, and Neil Armstrong examine the new XLR99 engine. (AFFTC History Office via the Gerald H. Balzer Collection)

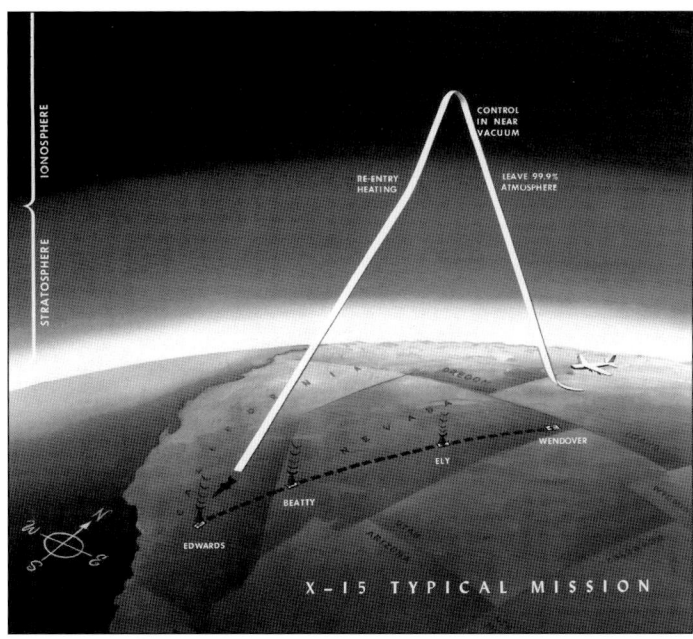

Typical mission profile showing the environmental and aerodynamic extremes controlling the design of the X-15. The drawing is a bit fanciful since no flights were ever launched from Wendover. (Boeing via the Gerald H. Balzer Collection)

The X-15 occupied most of the available room between the B-52's inboard engines and the fuselage. The swept wing of the B-52 created an asymmetrical downwash on the X-15 that caused it to roll as it was released. (AFFTC History Office via the Gerald H. Balzer Collection)

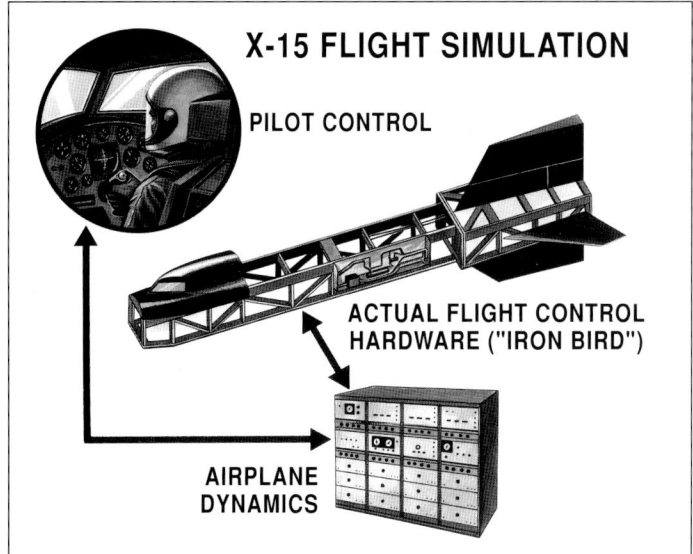

The simulator evolved considerably over the years. The original fixed-base simulator is shown at right as it existed on 16 September 1960. The drawing at left shows its configuration after the "iron bird" rig of flight hardware was relocated from North American. (NASA Dryden)

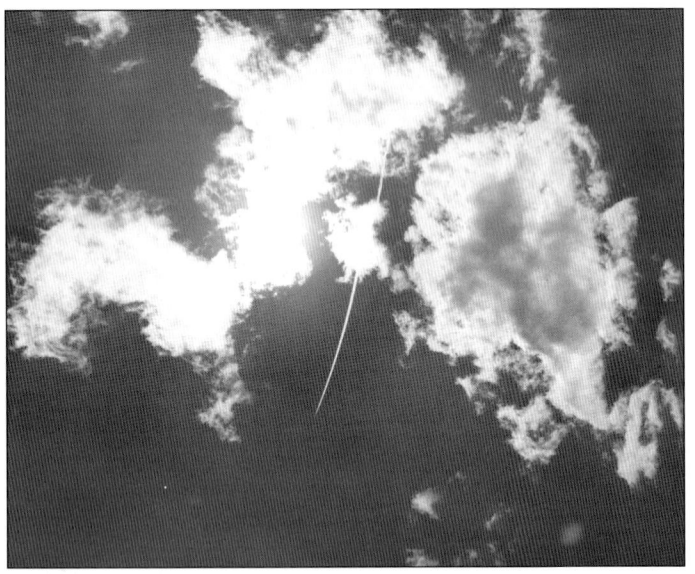

A ground view of flight 1-16-29 on 4 November 1960. (NASA Dryden)

Hoses connected to the jettison vents for servicing. (NASA Dryden)

The X-15 stable platform was tested using a pod suspended under Balls Eight. Here is the pod under construction on 30 November 1960; no photos could be found of the pod attached to the bomber, although at least seven flights were made during 1961. (NASA Dryden)

Unidentified test being conducted on an early example of the ball nose, sometimes incorrectly described as a "q ball" (the nose actually sensed alpha and beta). (NASA Dryden)

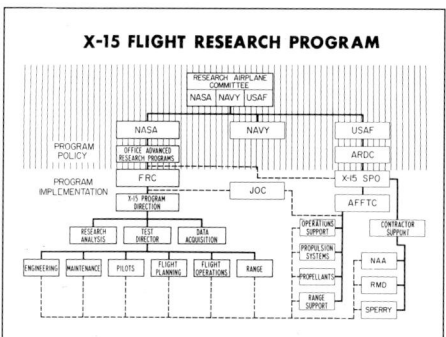

A test of the interim XLR11s at the Rocket Engine Test Facility. (Gerald H. Balzer Collection)

X-15 organization chart circa 1960. (NASA)

Landing sequence. Note the F-104 chase aircraft. Starfighters were almost always used for landing chase since they could easily fly the same landing profile as the research airplane. (AFFTC History Office via the Gerald H. Balzer Collection)

Things were more casual during the 1960s. Here is Bob Rushworth on 3 November 1965 walking by several vehicles on his way back from a flight. Note the X-15 program patch on the flight jacket of the officer at left. (U.S. Air Force via the Terry Panopalis Collection)

An X-15 mockup during a goodwill tour. The shape of the upper bug-eye camera fairings does not reflect how they were actually built, but otherwise the mockup appears to be accurate. (San Diego Aerospace Museum Collection)

Various temperature-sensitive paints were used during the X-15 program. These were painted onto a surface prior to a flight, then examined after landing. The paints turned various colors depending upon the highest temperature experienced during the flight. This is the lower surface of the left wing before (left) and after Flight 1-11-21 in August 1960. (NASA Dryden)

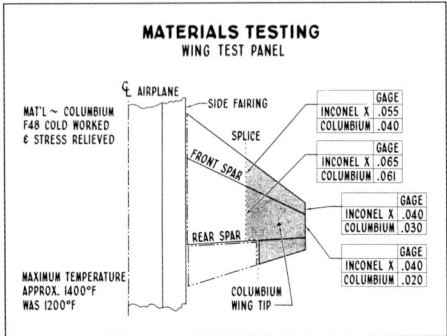 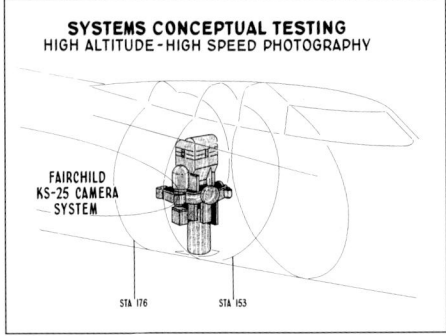 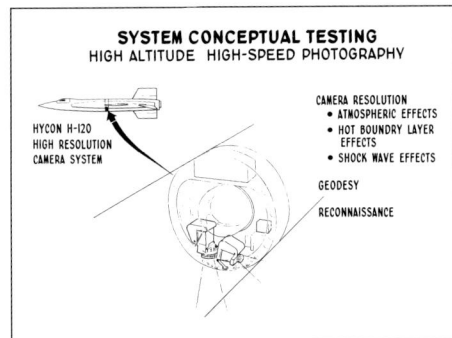

As early as 1960 North American had proposed using a removable wing tip to test various materials for future high-speed vehicles. The primary candidate material was columbium. (NASA)

A North American proposal to test a KS-25 reconnaissance camera at very high speeds was later implemented as the Photo Optical Degradation experiment (#5) during the Follow-On Program. (NASA)

Similarly, a proposal to install a Hycon camera in the center-of-gravity compartment was implemented on X-15-2 (and subsequently, X-15A-2) as the Hycon Camera experiment #27 during the Follow-On Program. (NASA)

X-15-2 heads for home after the Flight 2-13-26 landing on 7 March 1961. No expense was spared to procure a tow vehicle. (NASA Dryden)

Above and below: *Two views from the Edwards tower of Balls Three taxiing out on 19 December 1961. This flight attempt (the first for X-15-3) with Neil Armstrong would be aborted (3-A-1) when the XLR99 failed its pre-ignition checkout.* (AFFTC History Office)

Above: *Balls Three while it was still a JB-52A serving as a defensive platform development aircraft at Boeing Seattle. Note the gun turret in the extreme aft fuselage, and the external tanks under the outer wings.* (The Boeing Company Archives)

X-15 Photo Scrapbook

An often-used photo taken from the launch operator's astrodome in October 1961. Note the amount of flex in the NB-52 wing. The X-15-2 is equipped with upper bug-eye camera fairings and the ventral rudder. (National Geographic photograph by Dean Conger for NASA)

The X-15 ground crew in February 1961. Back row, from left: John "Bill" Lovett, John E. Huntington, Homer Hall, Robert E. "Bob" Allen, Lorenzo "Larry" Barnett, Charles "Charlie" Russell, and Sylvester Weeks. Kneeling: Gilbert "Gil" Kincaid, George E. Trott, Joseph "Joe" Huxman, Willard Glasscock. (NASA Dryden)

This photo of the NB-52B preparing for Flight 2-18-34 on 12 September 1962 is interesting primarily because it shows the transition from the large U.S. AIR FORCE marking on the forward fuselage to a smaller one (which has not yet been applied). Mission marks run the length of the forward fuselage. (AFFTC History Office)

Fight 2-19-35 falls away from Balls Eight on 28 September 1961. Note the slight variation in nose art on the NB-52B. (AFFTC History Office)

Flight 1-21-36 on 7 February 1961 was the last (and fastest) flight with the interim Reaction Motors XLR11 engines. (NASA Dryden)

Bob White (left) and Neil Armstrong walk away from the X-15-2. (AFFTC History Office via the Gerald H. Balzer Collection)

The X-15-2 in the Edwards weight and balance hangar with its ventral rudder installed. (Boeing via the Gerald H. Balzer Collection)

To determine how hard the surface of each dry lakebed was, engineers dropped an 18-pound steel ball and measured the resulting indentation. This is Hidden Hills in June 1958, with Neil Armstrong kneeling next to the ball. Note the markings on the R4D Gooney Bird. (NASA Dryden)

Bob Rushworth made the first flight (1-23-39) without a ventral rudder on 4 October 1961. Not installing the ventral rudder cured an instability at high angles of attack, providing a margin of safety should the stability augmentation system fail during reentry from high altitudes. The large vertical surfaces had originally been designed to cope with potential thrust misalignments that never materialized. (NASA Dryden)

The High Range station at Beatty on 30 December 1961. The Mod II radar antenna is on top of the building, with a microwave antenna facing away from the photographer at the left. The generator and fuel storage are in structures to the right. (NASA Dryden)

The NASA Gooney Bird at the airfield at Beatty, Nevada, on 13 December 1961. The commercial airfield was 5 miles on the other side of town from the High Range station, but was still less than 25 miles away by county road. (NASA Dryden)

The cutout in the wing of the NB-52 provided very little clearance for the X-15 vertical, but this never presented a serious problem during the flight program. (Boeing via the Gerald H. Balzer Collection)

Note that the upper surface of the X-15-1 wing is covered with a protective blanket during loading operations. (AFFTC History Office via the Gerald H. Balzer Collection)

X-15 Photo Scrapbook

A series of still images taken on 28 April 1961 during the making of the motion picture Project X-15. The main pilot character was played by David Mclean, the NASA Chief (in a tweed jacket) was James Gregory, the primary Air Force officer was Richard Kirby, and the X-15 backup pilot was Charles Bronson in his first motion picture. The X-15 replica used in the movie is now at Pima Air Museum representing the X-15A-2. Note square bug-eye camera fairings on the X-15. (NASA Dryden)

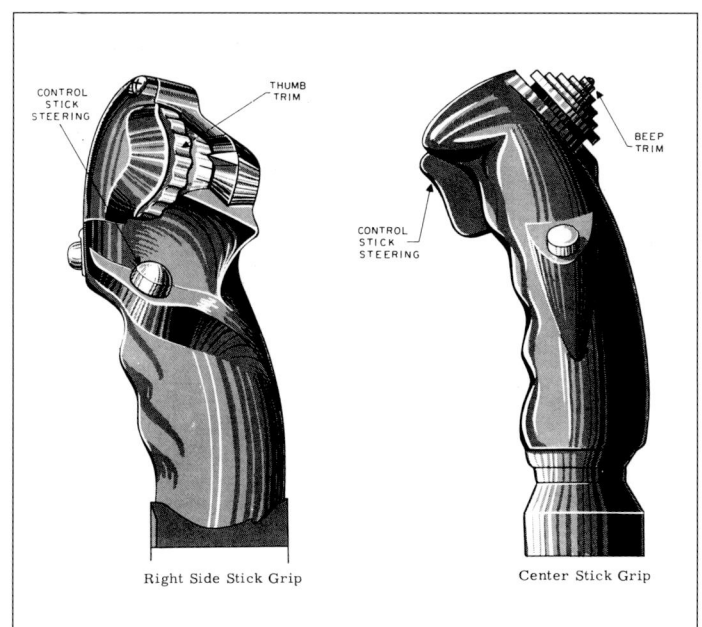

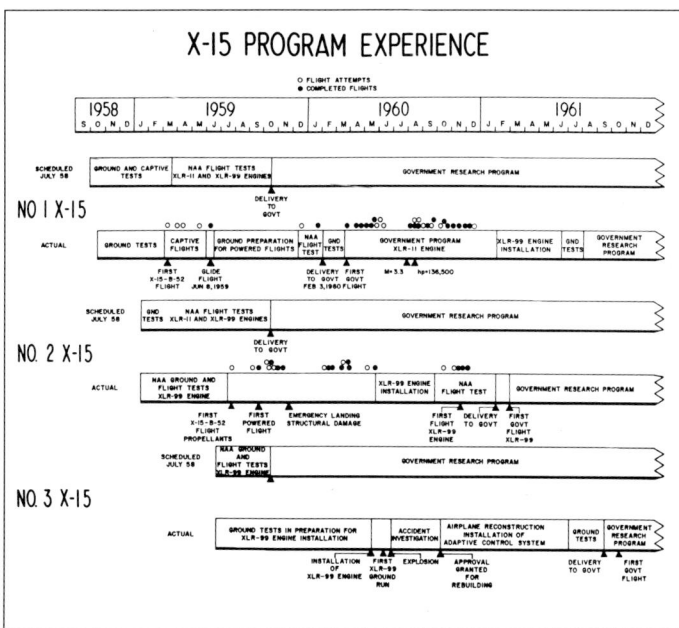

A couple of the switches for the MH-96 adaptive control system. With the control stick steering mode engaged, normal acceleration and pitch rate were fed back as the primary loop signals. Under this mode of operation, the system was designed to provide adequate damping and an essentially constant normal acceleration response. (Minneapolis-Honeywell)

A 1961 timeline showing significant events that had taken place during the X-15 flight program. For each airplane the upper bar represents the original July 1958 schedule, while the lower bar was when the events actually occurred. The first two airplanes were essentially on schedule, but the XLR99 explosion wrecked havoc with the X-15-3 schedule. (NASA)

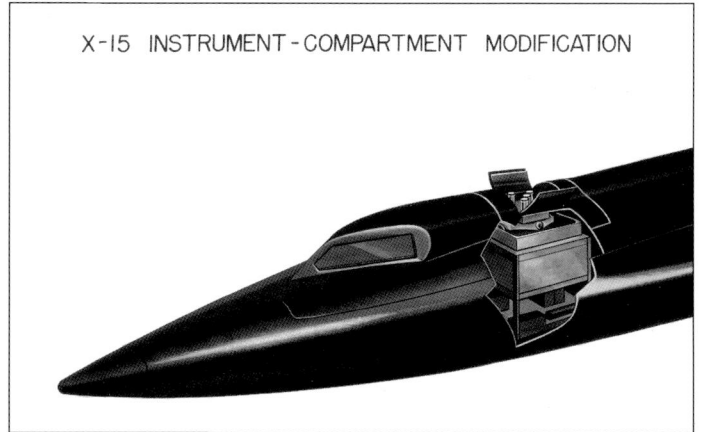

The "skylight" hatch concept was proposed as early as 1961, but its actual implementation was delayed after Jack McKay's accident with the X-15-2. Eventually slightly different skylight hatches would be installed in X-15-1 and X-15A-2. (NASA Dryden)

Forrest Petersen on 16 February 1961 next to the NASA JF-104A (56-0749) with an Air Launched Sounding Rocket (ALSOR). The intent was to release a balloon at 1,000,000 feet altitude and then measure its rate of descent to determine air density. (NASA Dryden)

The XLR-11 engine package from an X-15 during an open house on 21 November 1961. Note the X-15 has an XLR99 installed in it. (NASA Dryden)

The NASA-1 control room during Flight 2-21-37 on 9 November 1961. The concepts used in this area greatly influenced the mission control room for Mercury and Gemini. (NASA Dryden)

Typical scene during a ground test of X-15-3 on 12 April 1962. Except for engine runs, most testing was conducted in the NASA hangar, or on the ramp at the Flight Research Center. (NASA Dryden)

Engine runs, naturally, were conducted at the Rocket Engine Test Facility. Here is X-15-1 in the test stand on 1 February 1962. The concrete control bunker is visible at the left. (AFFTC History Office)

The X-15-3 on 15 January was fairly typical of the airplane configuration during 1962. Although it is difficult to discern in the head-on photo, the NASA tail stripe is only on one side of the airplane. The same photo shows only a single upper bug-eye camera fairing is installed. The photo below shows the antenna configuration and lower bug-eye camera fairings. (NASA Dryden)

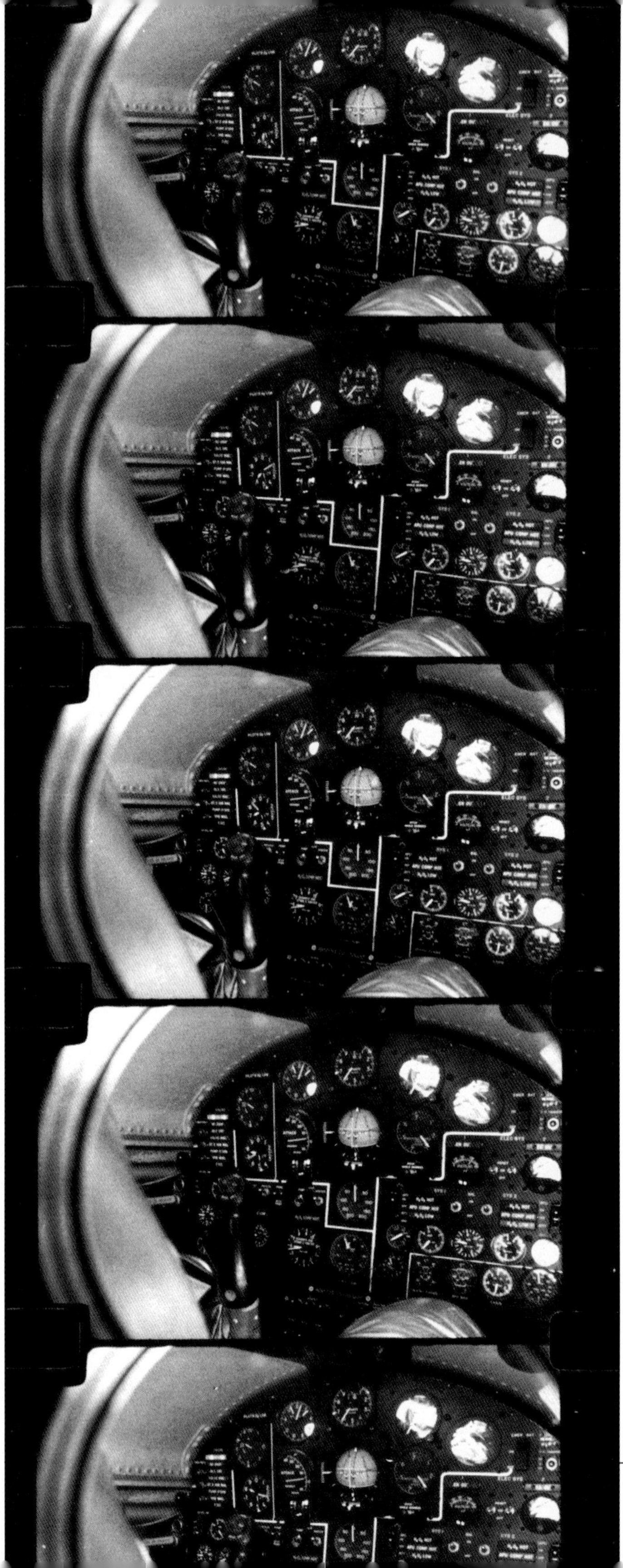

Left: *X-15-3 cockpit camera film from Flight 3-8-16 on 2 August 1962. Joe Walker and the airplane are at approximately Mach 3 and 55,000 feet.* (NASA Dryden)

Balls Eight and X-15-2 on 19 July 1962 for Flight 2-25-45. Note the small North American Aviation marking on the X-15 and the lack of a NASA Meatball. (AFFTC History Office)

This view of Flight 2-25-45 shows the NASA nose marking. Also note the large pipe on the side of the pylon that vented gaseous oxygen overboard. (AFFTC History Office)

X-15 Photo Scrapbook

The X-15-3 cockpit on 31 August 1962. At this point the X-15-3 did not look particularly different from the other two airplanes except for the MH-96 controls on the center pedestal and some warning lights. That would change when the advanced Lear-Seigler instrument panel was installed later in the flight program. Note the stopwatch used to control the engine burn time in the top center position. (NASA Dryden)

The wing leading edge expansion slots produced several problems during the flight program. It was soon learned that the expansion slots set up a wedge-shaped turbulent boundary layer behind them, significantly raising the heating rate on the wing skin. On one flight the area directly behind the expansion slots buckled. One reason was that the fastener spacing directly behind the slot was wider than on other sections of the leading edge, providing less support for the area. It was also determined that the original segmentation of the

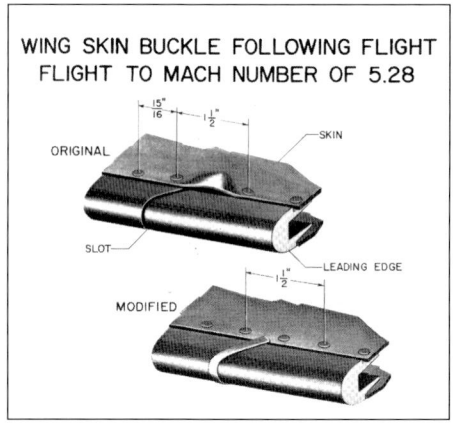
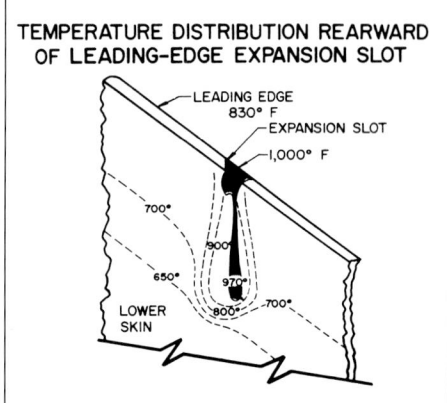

leading-edge heat sink did not adequately relieve the thermal-induced compression loads. The skins at the expansion slots acted as a splice plate for the solid heat-sink bar, and as a result buckled in compression. Several things were done to solve this: a shield was installed over each expansion slot to help the boundary layer problem and minimize the local hot spot, a fastener was added near each slot, and three additional expansion slots (with shields) were added in the outboard segments of the leading edge. (NASA)

An aerial view of Flight Research Center on 9 October 1962. Note X-15 and JF-104 on main ramp area. (NASA Dryden)

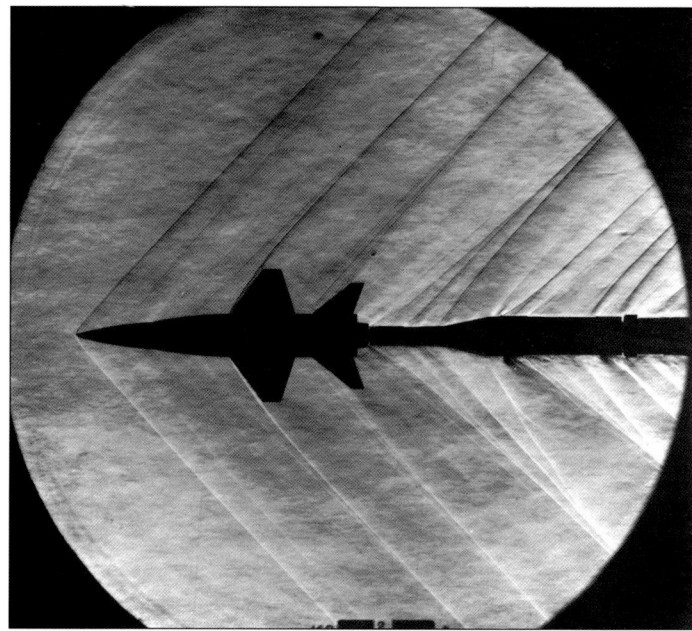

Shock waves festoon a small scale model of the X-15 in the Langley 4 x 4-foot Supersonic Pressure Tunnel on 23 March 1962. The X-15 underwent constant wind tunnel testing in a variety of facilities, although Langley conducted by far the most tests. (NASA Langley)

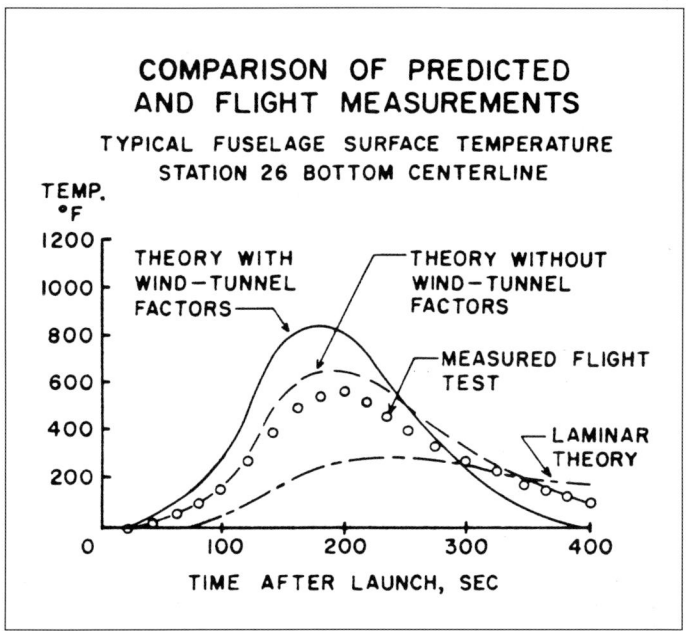

One of the primary reasons the X-15 existed was to validate various theories and wind tunnel testing methodologies against actual flight results. This chart shows the correlation between various methods as known in 1962. (NASA)

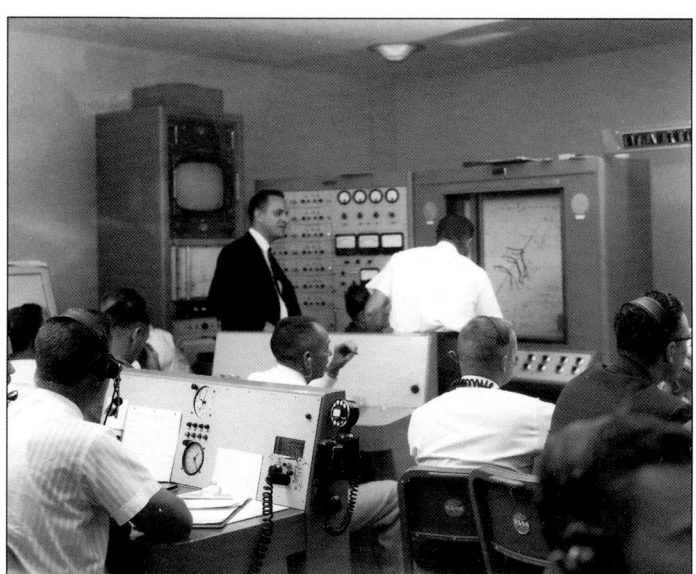

High Range operations during Flight 2-29-50 on 28 September 1962. (NASA Dryden)

Walter Cronkite interviews Jack McKay (in simulator cockpit) and Bob Rushworth for CBS News in October 1962. (NASA Dryden)

This and Facing Page: *The end of Flight 2-31-52 on 9 November 1962. As they went, Mud Lake was not a bad emergency landing site, but when the flaps failed Jack McKay's day turned from bad to worse. The left rear skid failed on touchdown and the airplane began to turn upside down. McKay jettisoned the canopy (below left) to ensure he could get out of the airplane after it stopped, but the result was his head hitting the lakebed, causing serious injuries. The rescue teams responded immediately, with the pilot of the H-21 helicopter hovering to disperse the ammonia fumes. McKay was airlifted to Edwards in the C-130 and eventually returned to make 22 more X-15 flights. Ultimately, however, his injuries forced his retirement from NASA and contributed to his untimely death.* (NASA Dryden)

X-15 Photo Scrapbook

Perhaps the most famous of the rocket pilots, Colonel Charles E. "Chuck" Yeager, looks at X-15-3 on 19 April 1963. Yeager would never fly the X-15, but surprisingly he did pilot the NB-52A for one flight attempt (2-A-38) on 25 April 1962. (NASA Dryden)

The crew of the X-15-3 on 9 August 1963 as Joe Walker prepares for the program's maximum altitude flight. Note nose art on X-15. Standing: Ernest R. Taylor, Harry ????, Febo Bartoli, and Joe Walker. Kneeling: Orville Cook, Unknown, James Rook. (NASA Dryden)

The second airplane being rebuilt into the X-15A-2 after Jack McKay's landing accident at Mud Lake. Note that most of the right wing is missing. (Boeing via the Gerald H. Balzer Collection)

An X-15-1 engine run on 19 August 1963. The XLR99 produced quite a flame, although – naturally – it was most impressive during test runs at night. (NASA Dryden)

A concerted effort was made to keep the instrument panels of the three airplanes as similar as possible, but differences in subsystems dictated they could not be identical. At left is the panel from X-15-1 on 26 January 1963, with X-15-3 shown at right on 23 April 1963. The major difference at this point was in the controls for the flight control system On the X-15-1 (and X-15-2) the Westinghouse stability augmentation system (SAS) and alternate Stability augmentation system (ASAS) controls were in the middle of the panel just behind the center stick. On X-15-3 this area was used for the controls for the Minneapolis-Honeywell MH-96 adaptive control system (which also performed the functions of the SAS/ASAS). Compare the X-15-3 panel with the earlier version shown on page 49. (NASA Dryden)

 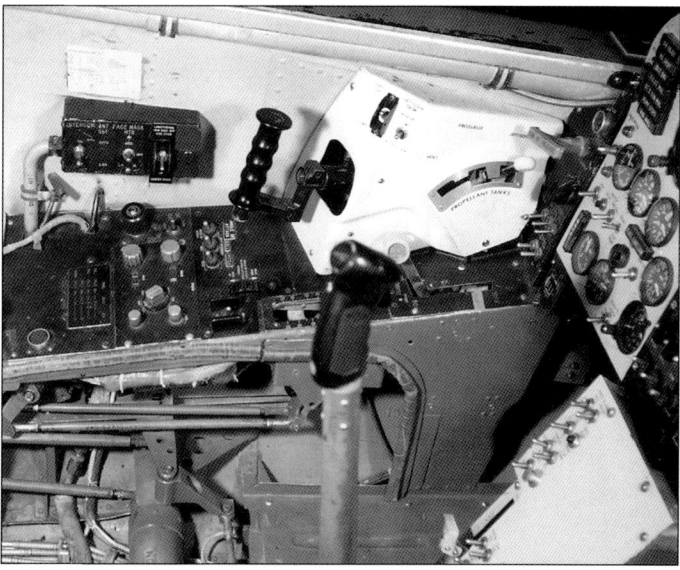

Despite the differences in the main instrument panels, the side consoles of all three aircraft were generally similar, although the X-15A-2 would add controls for the external tanks. These photos are of X-15-3 on 23 April 1962. (NASA Dryden)

X-15 Photo Scrapbook

Another angle of X-15-1 at the Rocket Engine Test Facility. Note the plastic sheeting surrounding the cockpit to provide some protection to the engineer in the cockpit in case of a propellant spill. The XLR99 could not be operated remotely, and a pilot or ground crewman had to be in the cockpit for ground runs. (NASA Dryden)

One of the more interesting follow-on experiments for the X-15 was to launch balloons from the tail-cone box in order to investigate the properties of supersonic wakes at high altitudes. This is the X-15-3 installation on 18 July 1963. The only two attempts to release balloons failed. (NASA via the Terry Panopalis Collection)

The X-15-1 at the Rocket Engine Test Facility on 19 September 1963. Note that most of the engine compartment covers and many of the side tunnel panels have been removed for maintenance. (NASA Dryden)

Each of the carrier aircraft painted mission markings on the nose. This is Balls Three on 1 August 1964 with the X-15-3 under the wing. Perhaps the most interesting mission marks are the pair above the X-15 BCS thrusters where one X-15 is pointing backwards; this was a ferry flight to Eglin AFB in May 1962 to show the X-15 to President Kennedy. (AFFTC History Office)

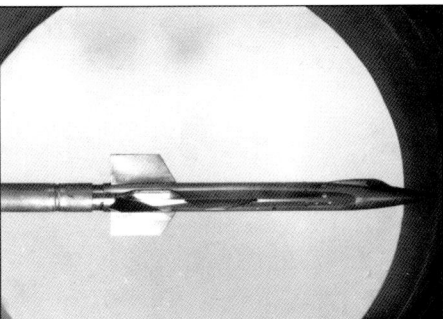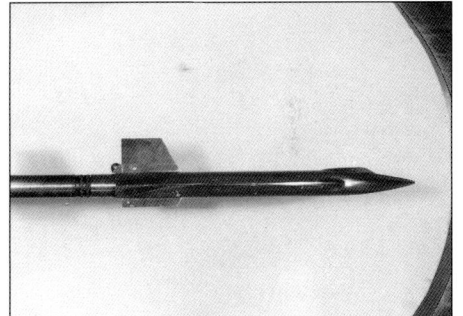

The Jet Propulsion laboratory in Pasadena was very involved in wind tunnel testing the X-15-A-2 configuration. The photo at left shows the advanced airplane with external tanks, and also with "slipper tanks" over the aft fuselage tunnels. The slipper tanks would have contained additional hydrogen (beyond what was to be carried internally) for the ramjet. The other two photos show variations in the vertical stabilizer that were evaluated in a search for additional stability at Mach 8. (Jet Propulsion Laboratory)

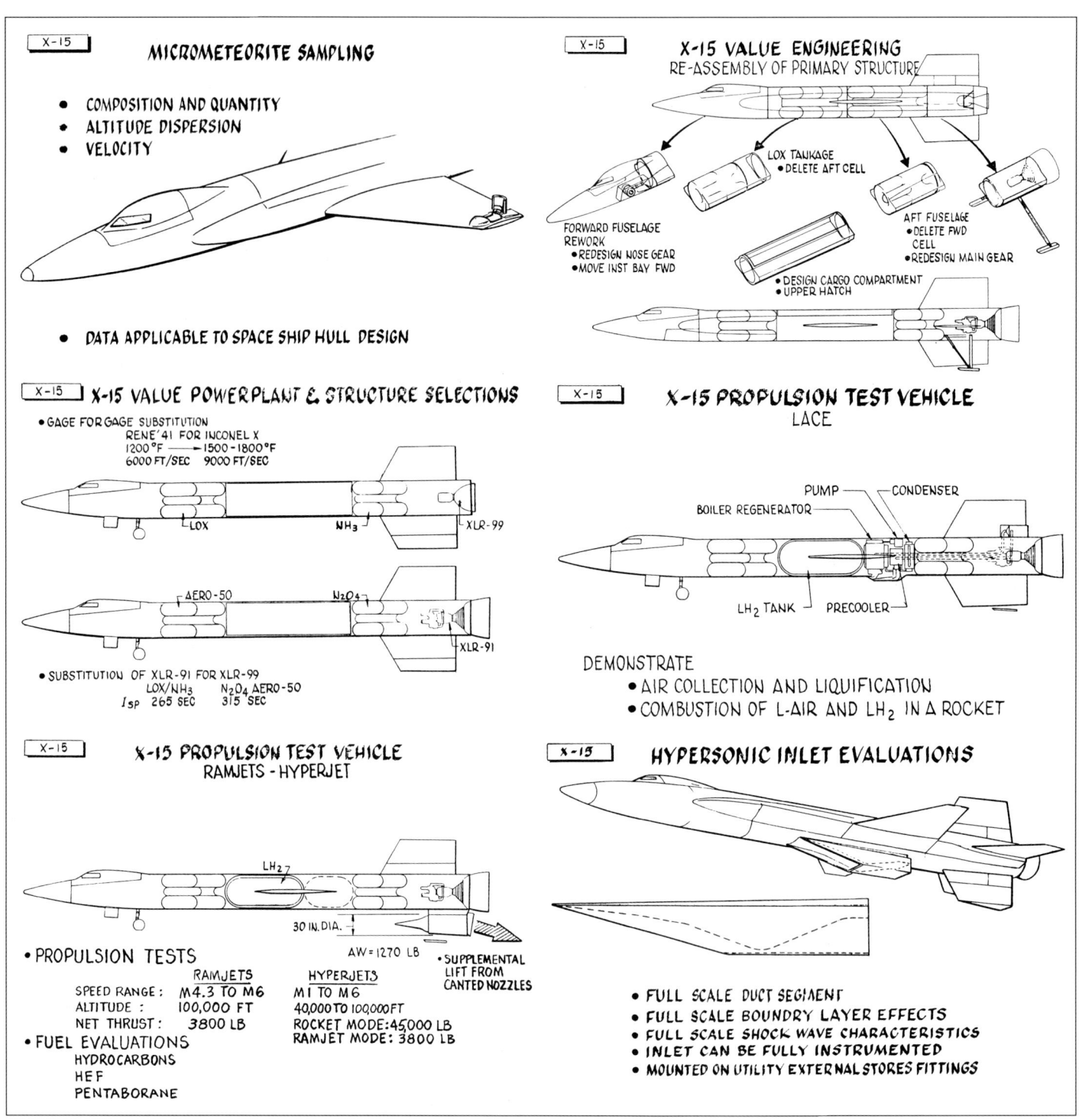

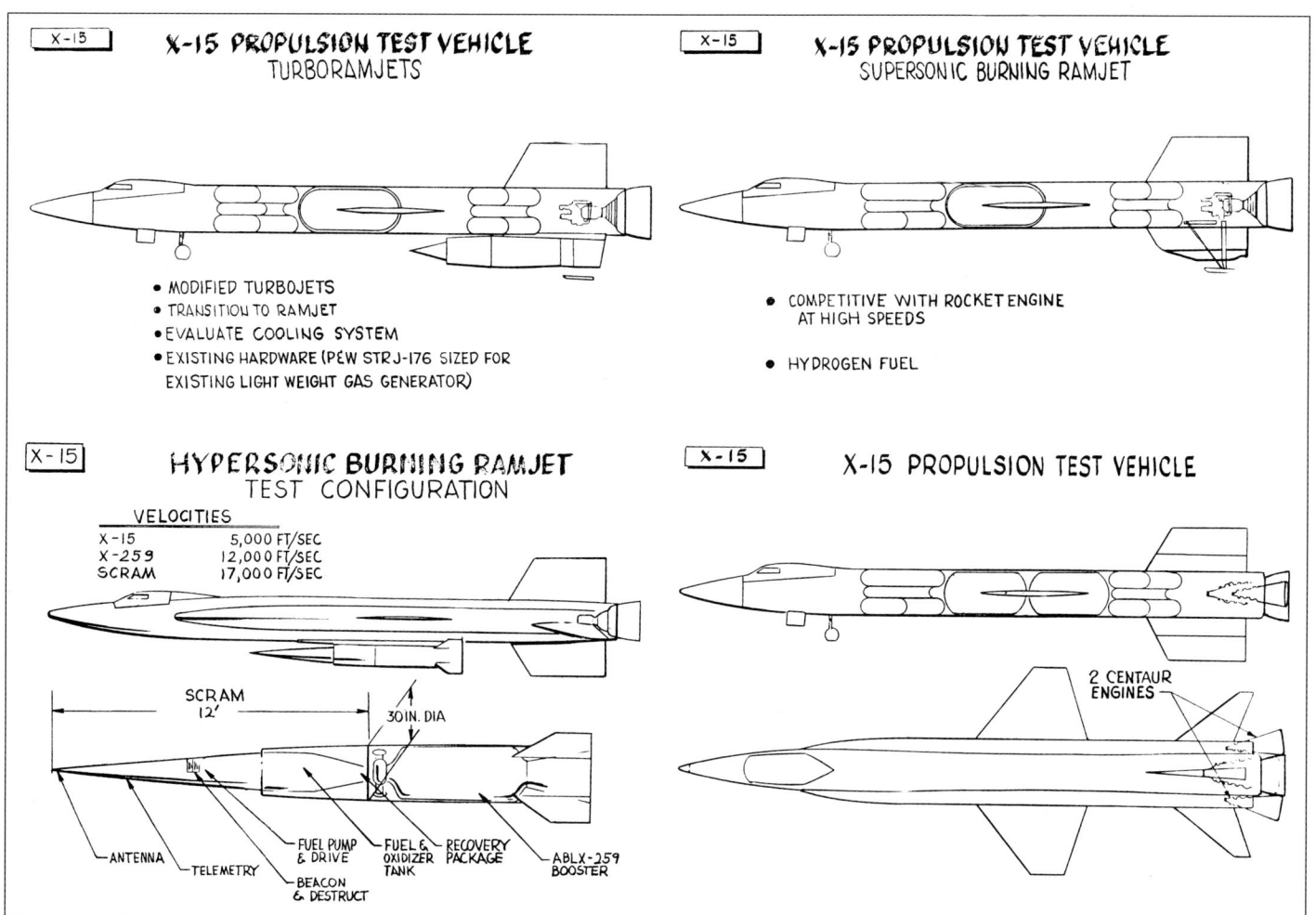

North American was continually proposing advanced missions for the X-15, and this set of briefing charts is from 1962 and concentrate on propulsion research. The micrometeorite collection experiment shown at the upper left on the facing page was actually implemented using a wing-tip pod collector, but this drawing shows one of the early proposals. Three of the drawings on the facing page show an X-15 modified with a central payload bay and upward-opening doors. This would have significantly reduced the propellant capacity of the design, and likely would have necessitated external tanks, although they are not shown on the drawings. The middle left drawing (facing page) shows the use of an Aerojet XLR91 engine that increased the available thrust by 70 percent. Note that most of the drawings show the nose landing gear moved aft into the instrument compartment. Various other engine combinations are shown in the remaining drawings, but none were actually flown. (The Boeing Company Archives)

X-15 Photo Scrapbook

The "advanced" X-15A-2 would finally provide the X-15 program the means to meet its original Mach 6.5 speed goal (the 250,000-foot altitude design goal had been easily exceeded several times). The X-15A-2 was rolled-out in Inglewood on 17 February 1964, and trucked to Edwards the following day. Although the airplane had been delivered to the Air Force at the roll-out, an official "acceptance" ceremony was held on 24 February. (NASA Dryden)

After going through several iterations of nose art, Balls Eight finally adopted this design and dropped The Challenger *name. This was photographed on 1 September 1964.* (NASA Dryden)

The Langley Horizon Definition experiment (#4) included a horizon scanner and 16mm camera installed in the tail-cone box of X-15-3 on 22 June 1964. (NASA Dryden)

The depleted silver material contained in the catalysts that turned hydrogen peroxide to steam for the XLR99 turbopump, APUs, and ballistic control system proved to be very handy for making gifts for employees and VIPs. (NASA Dryden)

President Lyndon B. Johnson visited Edwards on 19 June 1964, and reviewed the X-15-3 and X-15A-2 parked on the ramp. More photos may be found in the color section. (Terry Panopalis Collection)

A variety of different ramjet/scramjet/rocket engines were proposed for testing on the X-15 during the course of the program. Many mockup installations were even test-fitted, such as this hybrid ramjet/rocket engine on X-15-3 on 3 June 1964. None of the engines were ever actually flown, except the dummy ramjet flown on X-15A-2, which was part of the Hypersonic Research Engine (HRE) program. (NASA Dryden)

Another proposed engine installation, this time a scramjet under the ventral of the X-15A-2 on 19 June 1964. (NASA Dryden)

Bob Rushworth had an odd string of problems when the landing gear began deploying at high Mach numbers. It all started during Flight 2-33-56 on 14 August 1964 when the nose gear came down as Rushworth decelerated through Mach 4.2 – needless to say, the tires were not worth much during the landing. Fortunately the wheels and nose strut were not seriously damaged and the flight concluded without major incident. The culprit was aerodynamic heating: the fuselage "grew" more than the slack in the landing gear release cable could compensate for, resulting in a "pull" that released the uplocks. (NASA Dryden)

As a result of Rushworth's experience in August, the Flight Research Center began a series of nose gear heating tests on X-15A-2 on 2 September 1964. These eventually led to a lengthening of the release cable as well as some other minor fixes. (NASA Dryden)

Both X-15-1 and X-15-3 were equipped with wing-tip pods to carry experiments for the Follow-On Program. This is X-15-1 on 15 October 1964 after Flight 1-50-79 with Jack McKay at the controls. Note the cover over the ball nose. (NASA Dryden)

X-15 Photo Scrapbook

For reasons that could not be ascertained, the ball nose was tested on a Douglas F5D Skylancer during October 1964. The installation looked a bit odd, and no record could be uncovered regarding how many times the unit was flown in this configuration. (NASA Dryden)

A simple expedient to testing the ball nose under high-heat conditions was to place the unit in the afterburner flame. This was done using an F-100 for the early ball noses, and a J79 from an F-104 for the TAZ-8A nose for the X-15A-2 (shown above). (NASA Dryden)

X-15-3 at Cuddeback Lake after Milt Thompson had to make an emergency landing during Flight 3-29-48 on 21 May 1964. This was the first of two landings at Cuddeback. (NASA Dryden)

Above and left: *One of the few times all three X-15s were photographed together was on 23 April 1965 during the filming of Project X-15. In the photo above they are lined up in serial number order, with X-15-1 in the foreground and X-15-3 (note the sharp leading edge on the dorsal rudder) in the background.* (NASA Dryden)

Preflight servicing of X-15-1 on 15 October 1964. Note the tail-cone box and wing-tip pods. (NASA Dryden)

Various ablators were tested on the X-15 during the course of the flight program. Some of these were directly applicable to the X-15A-2, while others were being evaluated for the Saturn launch vehicle or other applications. This photo shows a sample of General Electric ablative material on the fixed ventral of X-15-3 on 16 July 1964. There was a period of time during the 1960s and very early 1970s were ablative materials were being considered for what became the Space Shuttle. The mostly unhappy experience with the full-scale ablator on the maximum speed flight of the X-15A-2 ended all thoughts of using it on larger vehicles. (NASA Dryden)

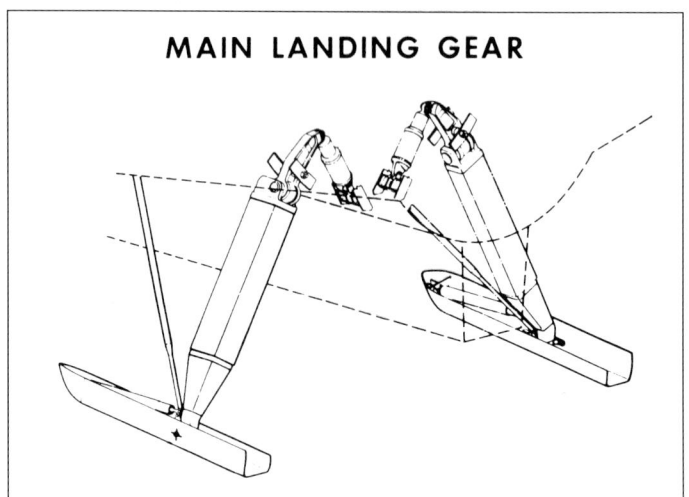

The main landing gear on the X-15 was a study in simplicity. A skid on each side, carried externally to the airframe. The release mechanism was a cable that was pulled by the pilot to release the uplocks, deploying the skids and nose gear. (NASA Dryden)

Landing gear loads were a constant concern to the X-15 program, mainly because the airplanes were heavier than expected and constantly got heavier as new equipment was added. This is the main landing gear test setup on 23 April 1964. (NASA Dryden)

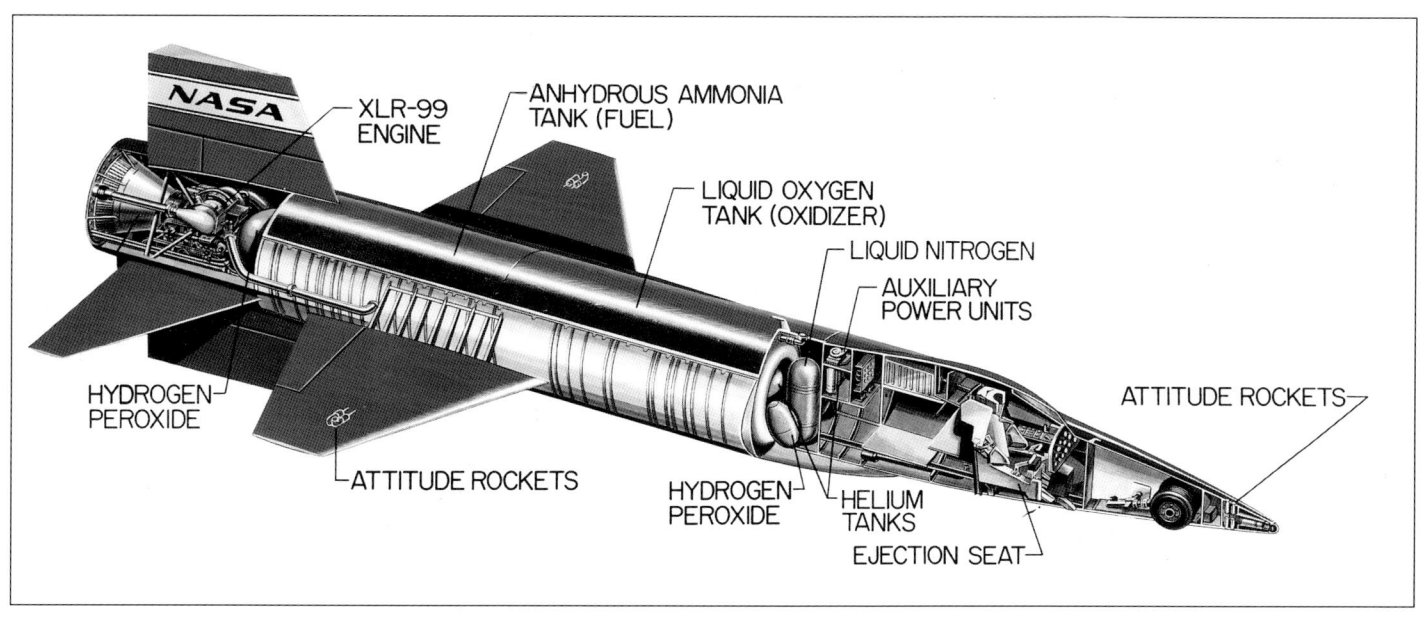

The X-15 was essentially a flying propellant tank, with a rocket engine in the back and space for a pilot and research instrumentation in the front. Most of the structure and all of the skin was made from Inconel X, essentially a heat-resistant stainless steel. (NASA)

The X-15-3 in the paint shop at the Flight Research Center on 25 January 1965. Note the cabling coming out of the instrument compartment behind the cockpit. (NASA Dryden)

Another ablator being tested, this time a Martin product on the fixed portion of the vertical during Flight 3-41-64 on 16 April 1965. Note the sharp leading edge on the dorsal rudder. (NASA Dryden)

X-15 Photo Scrapbook

This scanner and radiometer in the tail-cone box of X-15-3 on 27 May 1965 was part of the Langley Horizon Definition experiment (#4) in the Follow-On Program. This experiment was intended to examine the visible and far-infrared spectrum to determine if certain horizon phenomena were sufficiently stable to serve as Apollo space-navigation references (they were). (NASA Dryden)

This view of the X-15A-2 on 8 October 1965 emphasizes how large the external tanks were. These tanks allowed engine run times to essentially double, providing a theoretical increase in speed from the Mach 6 recorded by the basic airplane, to almost Mach 8. Unfortunately, the program ended at Mach 6.7 after encountering difficulties with the HRE ramjet and MA-25S ablator. (NASA Dryden)

The Inconel X skin was capable of withstanding high temperatures, but it still showed the effects. These are wrinkles in the forward fuselage of X-15-3 on 8 February 1965. (NASA Dryden)

Much of the instrumentation was "added on" as needed to support research objectives. This is a shear layer rake on the vertical stabilizer of X-15-3 on 3 June 1965. (NASA Dryden)

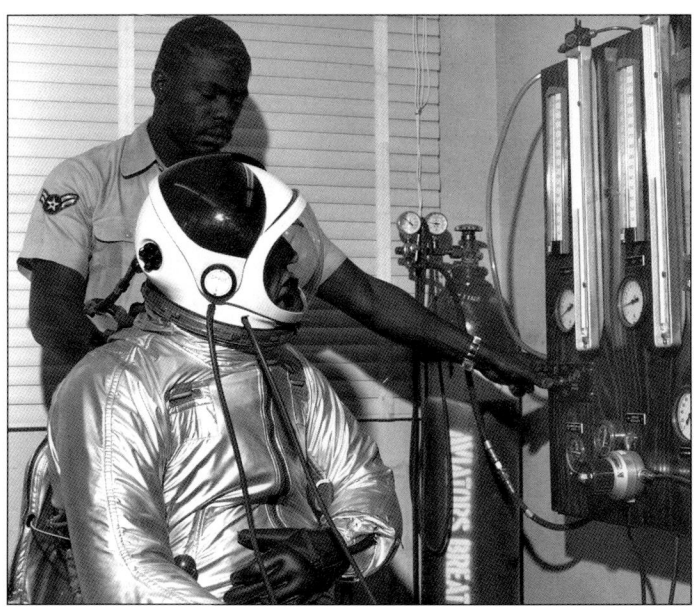

Captain Ralph N. Richardson, the chief of the physiological support branch at the AFFTC, is shown wearing a David Clark Company A/P22-S full-pressure suit. This suit was a significant improvement – in ease of use, comfort, and visibility – over the MC-2 suits used for 36 early flights. (AFFTC via the Gerald H. Balzer Collection)

Getting ready for Flight 1-52-85 on 26 February 1965. Note the wing-tip pod on the X-15-1. The camera window on the bottom of the fuselage had been installed for the Photo Optical Degradation experiment (#5), but was being used for the Infrared Scanning Radiometer (#23) when this photo was taken. (NASA Dryden)

Part of clearing the external tanks for flight was a series of jettison tests conducted with the X-15A-2 suspended over a pit dug beside the ramp. These used a steel beam that had the same weight as the tanks. The photo was taken on 29 September 1965. (NASA Dryden)

The X-15-3 under the wing of Balls Eight for Flight 3-50-74 on 12 October 1965. By this time the carrier aircraft had lost their colorful international orange markings and were overall silver. Fortunately, the nose art and mission marks remained. (AFFTC History Office)

X-15 Photo Scrapbook

A crew from the Flight Research Center building the X-15 BCS display for the 1965 X-15 industry conference. Moving the side-stick controller would fire the BCS thrusters on a model that would be located just in front of the cockpit (see below). (NASA Dryden)

Joe Engle in his A/P22S-2 pressure suit after Flight 3-44-67 on 29 June 1965. Engle would later have the distinction of being the only person to manual fly significant portions of a Space Shuttle reentry. (AFFTC History Office)

The X-15 program held four industry conferences to disseminate results from the research. The last of these conferences was held at the Flight Research Center on 7 October 1965 and featured various displays including the BCS display and X-15A-2. (NASA Dryden)

Bob Rushworth and Joe Engle show off their astronaut wings on 5 August 1965. The military pilots qualified as astronauts if they flew above 50 miles. NASA used the international 62-mile standard, but even then, Joe Walker never received the recognition. (NASA Dryden)

The speed brakes on the upper and lower vertical stabilizers opened to form an included-angle of 35 degrees. Normally both speed brakes were operated in unison, but at least one flight experimented with operating them independently. (San Diego Aerospace Museum Collection)

Not quite what the public relations department wanted. Joe Engle, Milt Thompson, Bob Rushworth, Pete Knight, Jack McKay, and Bill Dana clown around for the camera on 2 December 1965. Its a serious business, but you need to maintain your sense of humor – something most of the pilots were very good at. (NASA Dryden)

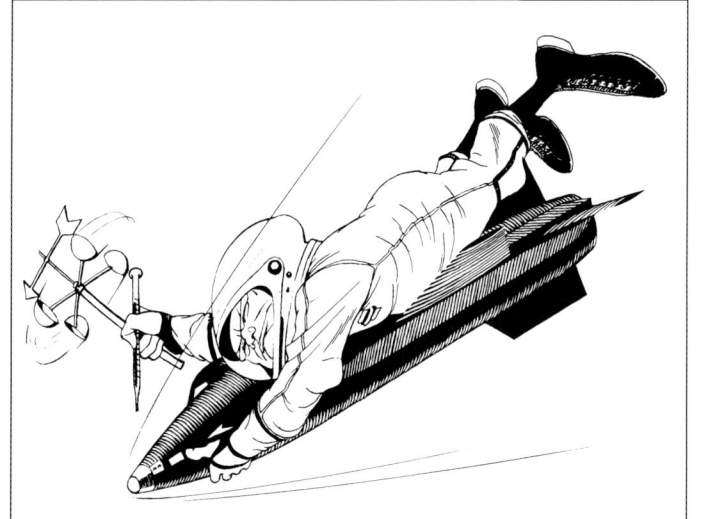

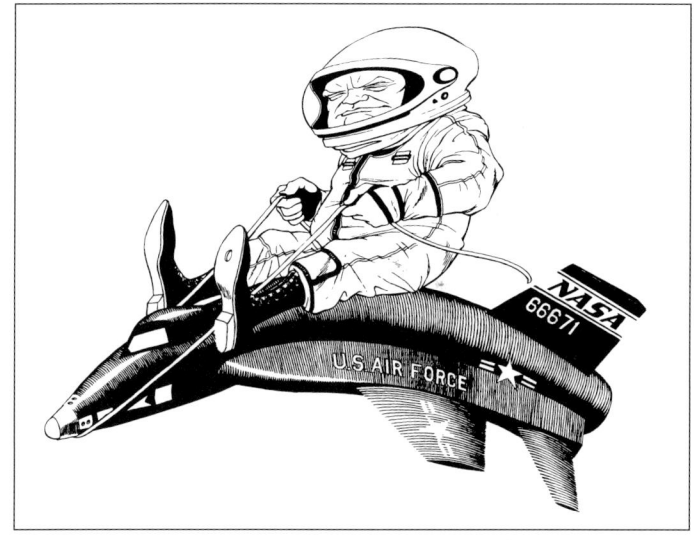

Jack McKay flew the X-15 29 times, more than anybody except Bob Rushworth; he had flown other rocket-planes 46 times, accumulating more rocket flights than any U.S. pilot except Scott Crossfield. McKay also has the distinction of landing the X-15 at more different lakes than anybody else: of his 29 flights, twenty-six landed at Rogers, one at Mud, one at Delamar, and one at Smith Ranch. The one at Mud almost killed him, and ultimately contributed to his early retirement and untimely death. (NASA Dryden)

Temperature-sensitive DetectoTemp paint reveals the internal structure of the X-15-3 dorsal rudder on 14 September 1966. Note the sharp leading edge modification and the two expansion slots. (NASA Dryden)

Balls Three heading out for Flight 3-54-80 on 19 August 1966. Note the ventral rudder is missing from the X-15-3. The carrier aircraft pilots usually taxied with the right outrigger gear running on the ground to balance the weight of the X-15. (AFFTC History Office)

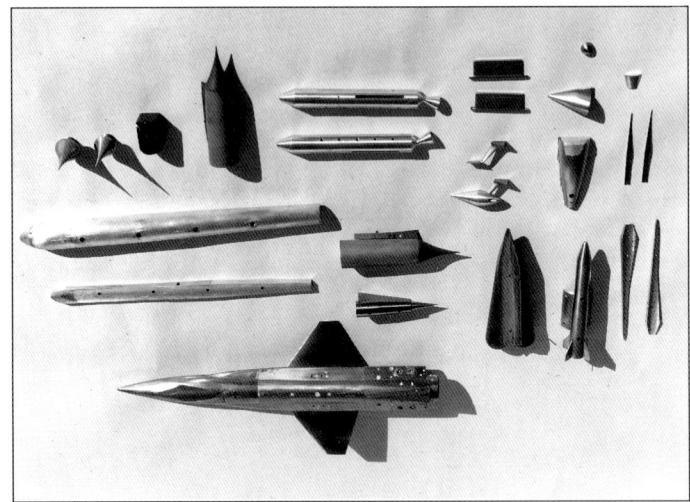

An X-15 wind tunnel model and add-on various components on 22 January 1965. Note the pair of rocket boosters (top center), the Blue Scout sounding rocket (lower right), and the various ramjet/scramjet shapes. (Tony Landis Collection)

Ground runs of the XLR99 in the airplane were conducted before every flight during the first part of the program. Later, if the engine had not caused problems on a flight, the ground run was often skipped. This is X-15-3 on 31 September 1966. (NASA Dryden)

 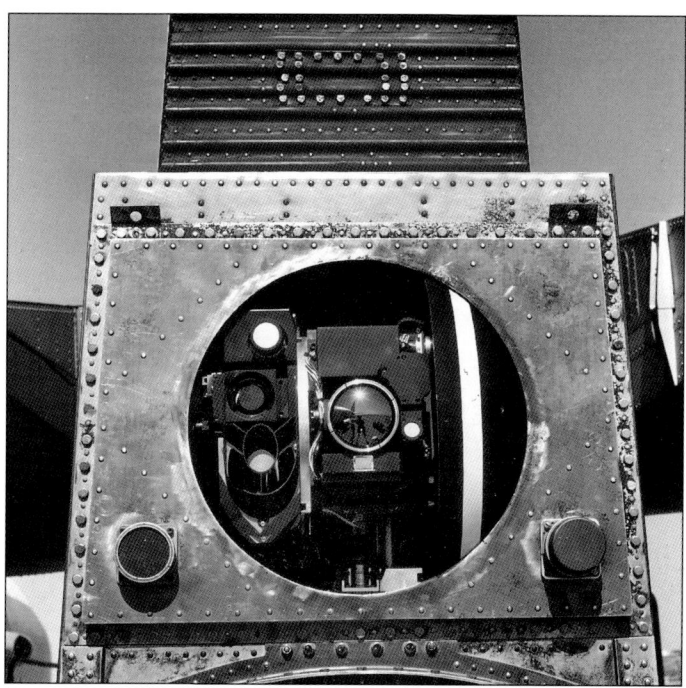

Several of the tailcone boxes had doors that covered the experiment during the boost phase so that exhaust efflux did not contaminate the instruments. This is the MIT-Apollo Horizon Photometer experiment (#17) on 28 April 1968 installed in X-15-1 for Flight 1-76-134. The circular door pivoted to one side of the photometer and camera as shown in the photo at right. (NASA Dryden)

Scenes from the Mud Lake recovery operations following Flight 2-45-81. The NASA Gooney Bird (R4D) and an AFFTC C-130 ferried men and equipment up to the lake. The dorsal rudder was removed and the X-15A-2 was placed on a special trailer (at right) for the trip back to Edwards. (AFFTC History Office)

X-15 Photo Scrapbook

Pete Knight (left) and Bill Dana during an interview on 1 November 1966. Note the large X-15 model in the background. (NASA Dryden)

Two of the X-15s being serviced in the NASA hangar – neither has a visible serial number. (NASA via the Gerald H. Balzer Collection)

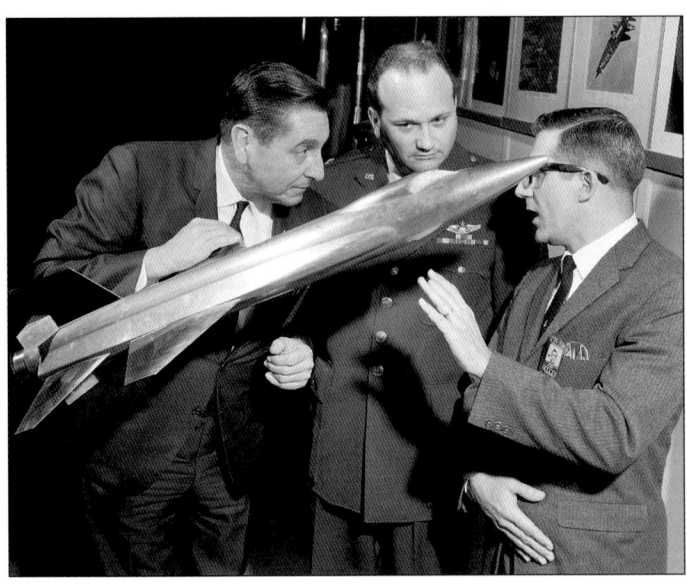
Scott Crossfield (left) had been gone from the program for several years at the time of this 1966 photo, but nevertheless poses with Chuck Schuler (right) and Major McCready from the Air Force Arnold Engineering Development Center. (AEDC)

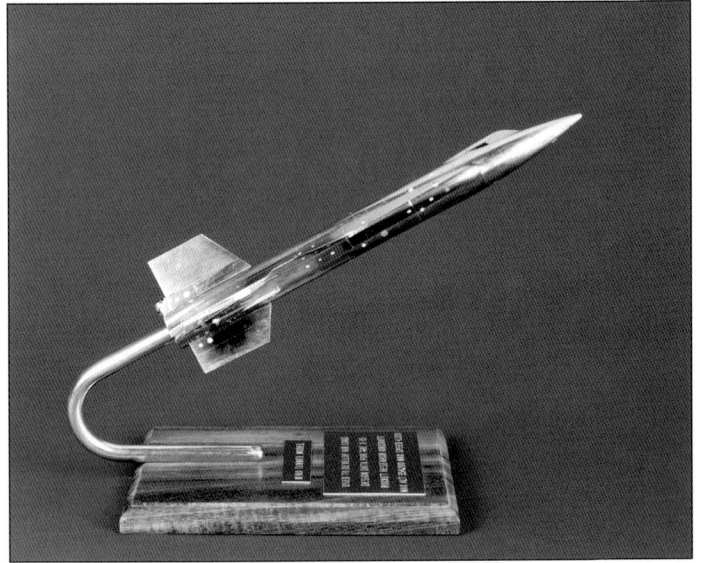
Wind tunnel model of the X-15A-2 with a proposed 60/40 vertical stabilizer arrangement. The concept was never applied to the actual airplane. This model was presented to Bill Dana upon his retirement from NASA. Photographed on 11 June 2002. (Tony Landis)

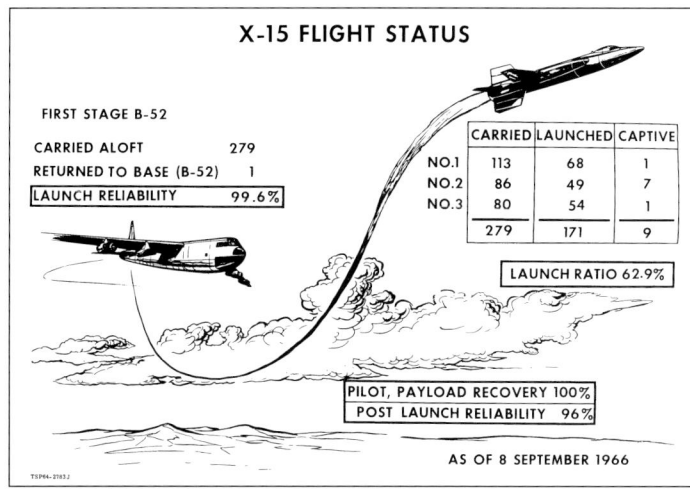

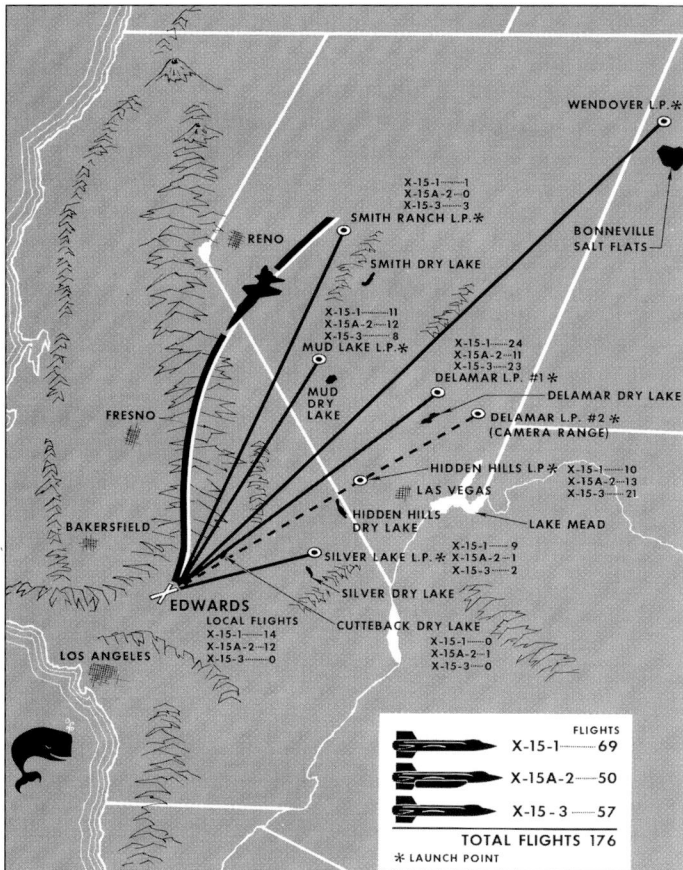

Two charts showing launches-to-date for the X-15 program. The top one gives a percentage of successful launches versus attempts, while the lower one shows the basic launch corridors used by the program. (Gerald H. Balzer Collection)

The most ambitious of the Follow-On experiments was the Western Test Range Launch Monitoring experiment (#20) installed in X-15-1, shown here on 2 August 1966. The X-15 was supposed to track a Minuteman ICBM launched from Vandenberg AFB. (NASA Dryden)

Jack McKay made an emergency landing at Delamar Lake on 6 May 1966 when the XLR99 decided to be problematic. The aircraft slid off the end of the lakebed, but sustained little damage. McKay, however, did jettison the canopy as a safety precaution. During the emergency landing, dust was blown into the Atmospheric Density Measurement experiment (#12) in the right wing-tip pod and acted as an abrasive on the radioactive material used in the experiment. The situation did not create a hazard from the radioactivity; however, the experiment was damaged and had to be returned to its researcher for repairs. (NASA Dryden)

This particular scramjet-powered X-15 completely replaced the XLR99 with a scramjet propulsion package that occupied the normal ammonia tank location as well as the engine compartment. The design never got beyond this model and some minor engineering evaluations. (The Boeing Company Archives)

The dorsal rudder of X-15-3 was modified with a "sharp" leading edge instead of the more rounded one normally carried on the X-15 (referring to the radius of the curvature of the leading edge). The leading edge was also extended forward several inches as shown in the 4 August 1966 photo. (NASA via the Terry Panopalis Collection)

An artist concept of a very heavily modified delta-wing X-15 with two large ramjets being launched from a B-70 carrier aircraft. The idea of using a B-70 was a recurring concept during the course of the program, but never happened. (Tony Landis Collection)

The X-15-3 vertical stabilizer with the cold wall heat transfer experiment equipment installed on 18 April 1967. The lighter colored rectangular section of skin could be jettisoned at high speeds, exposing thermocouples to the sudden heat. (NASA Dryden)

X-15 Photo Scrapbook

Two high-speed North American aircraft pose together on 4 August 1967. The XB-70A Valkyrie could "only" fly at Mach 3, but weighed well in excess of 500,000 pounds. The Mach 6-plus X-15A-2 weighed only 50,000 pounds. (NASA Dryden)

The full-scale ablator was applied in June 1967, with the airplane being rolled out of the paint shop on the 21st, resplendent in its bright pink Martin MA-25S. Subsequently, a white wear layer of Dow Corning DC-90-090 RTV was applied. *(NASA Dryden)*

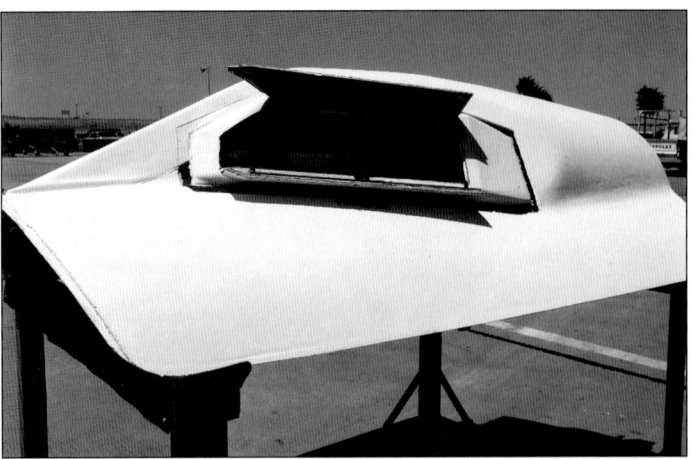

A couple of views of the canopy used on the X-15A-2 after the eyelid was installed. The eyelid was meant to protect the window from being smeared by ablator residue during flight and was opened at about Mach 1.5 to provide visibility during landing. (NASA Dryden)

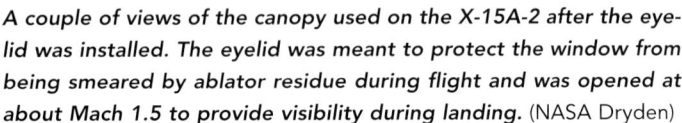

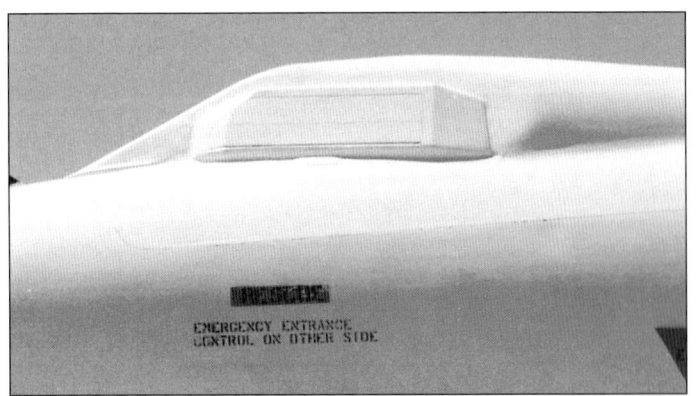

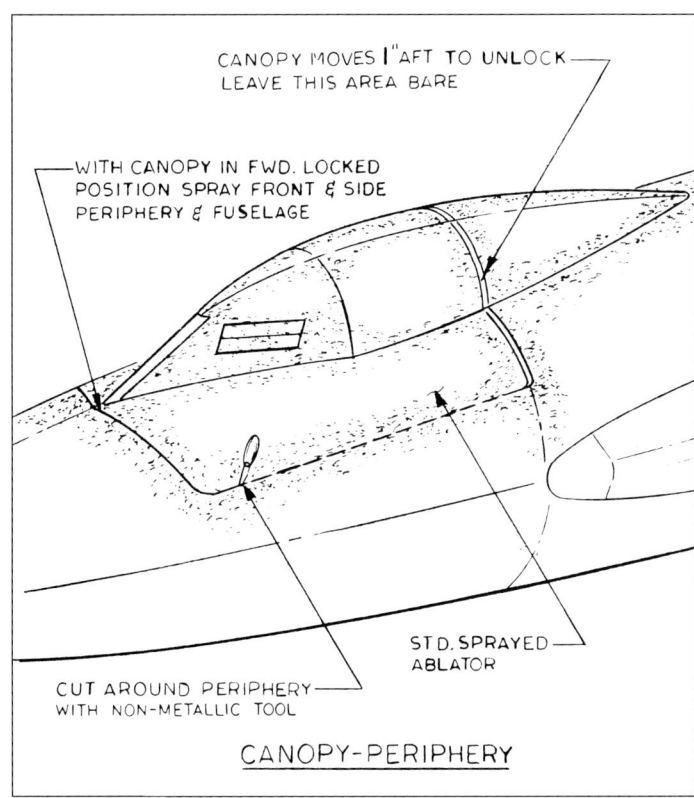

One of the illustrations from the instructions for applying the ablator on X-15A-2. Of interest is the note on how the canopy slides aft to open. (The Martin Company)

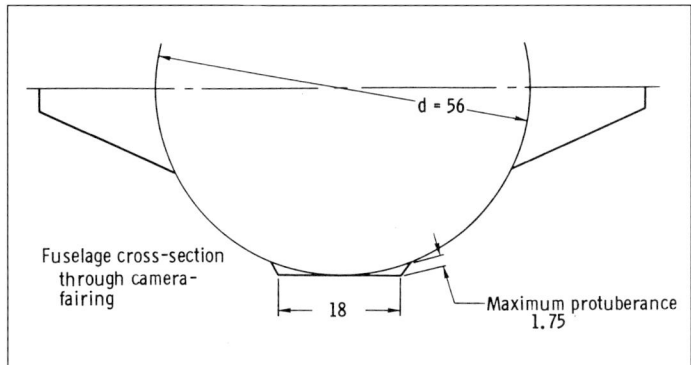

The X-15A-2 carried a Hycon camera in the center-of-gravity compartment between the propellant tanks. This drawing shows the camera window on the bottom of the fuselage. When the ramjet was installed the Hycon was replaced by a 16mm movie camera. (NASA)

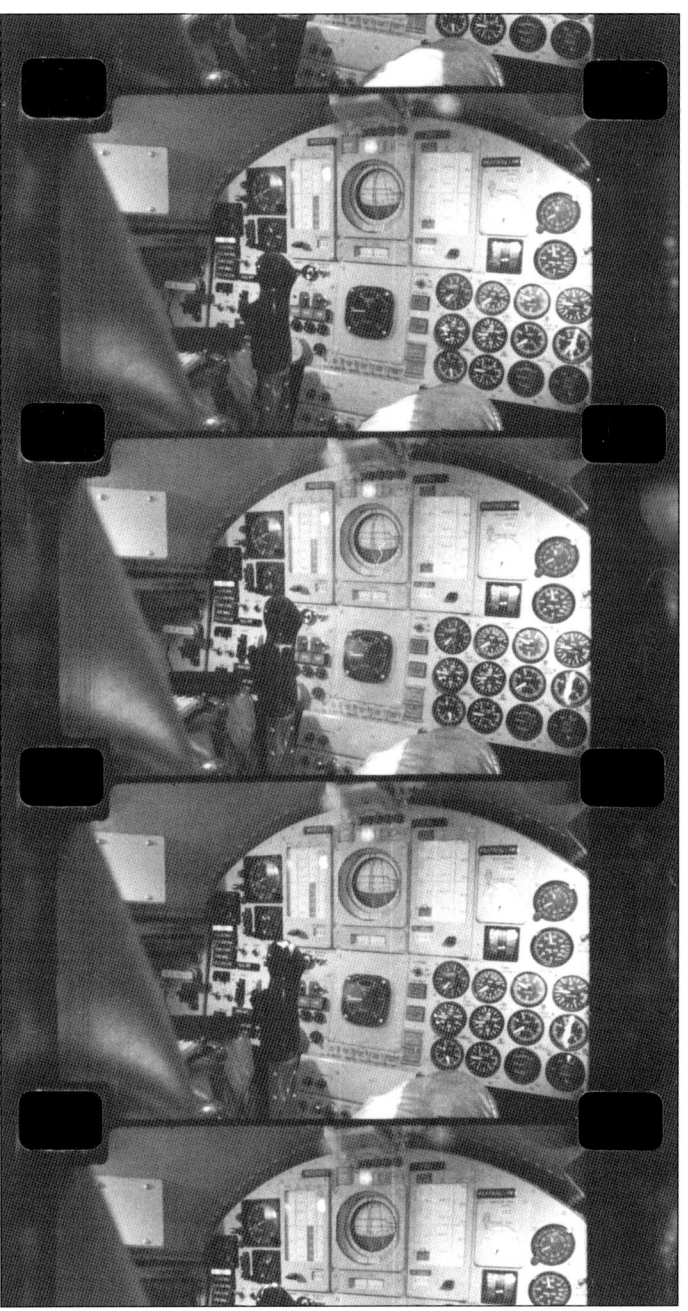

Cockpit film from Flight 3-52-78 on 18 July 1966. Although the advanced Lear-Seigler instrument panel has been installed, the Ames Boost Guidance Display in the lower center part of the panel is missing, replaced by an unidentified analog instrument. (NASA Dryden)

X-15 Photo Scrapbook

We do not fully understand this photo either. The object is apparently a ventral rudder, but exactly what the structure is around it is unknown. It is possible this unit was drop tested from a helicopter to verify the parachute system. (NASA Dryden)

Abort 2-A-95 on 16 August 1967 was caused by a pressure loss in APU #2. By This time Northrop T-38A Talons had taken over as Chase-1 from the original North American F-100 Super Sabres. (AFFTC History Office)

A loads test being conducted on the horizontal stabilizer of X-15-3 on 15 August 1967. It was not unusual for the X-15s to be tested on the ground for comparison against data gathered during flight. For most of the program the horizontal stabilizers were not instrumented because they were so thin. Later a set of instrumented units was installed for a few flights of X-15-3. (NASA Dryden)

The test of insulation for the Saturn launch vehicles is usually heralded as one of the X-15s contributions, but in reality the tests were were very minimal and concentrated more on the adhesives behind the insulation. Here the Saturn insulation is installed on the upper speed brakes of X-15-3 on 31 October 1967. Note the tail cone box behind the speed brakes. (NASA Dryden)

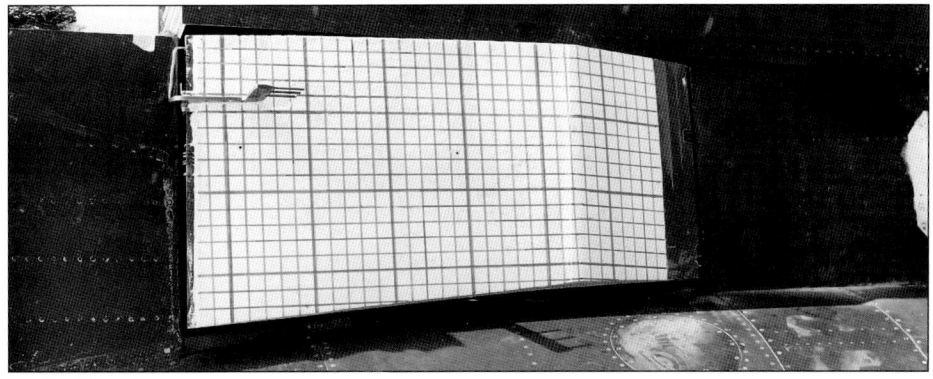

Insulation for the second stage of the Saturn launch vehicle on the speed brakes of Flight 1-76-134. The grid pattern was unusual, and may only have flown once. The photo below shows the insulation at the end of the Mach 5.05 flight on 26 April 1968. (NASA Dryden)

The photos on either side show a large-scale ablator being applied to X-15-1, but no record can be found of the airplane ever flying in that configuration. It is possible that the X-15-1 was a convenient stand-in for the X-15A-2 to allow the Martin technicians to practice applying the ablator. (NASA Dryden)

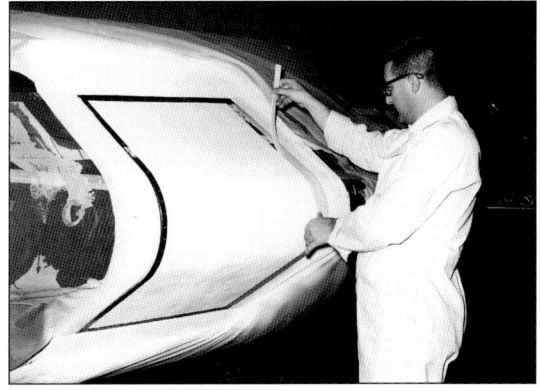

X-15 Photo Scrapbook

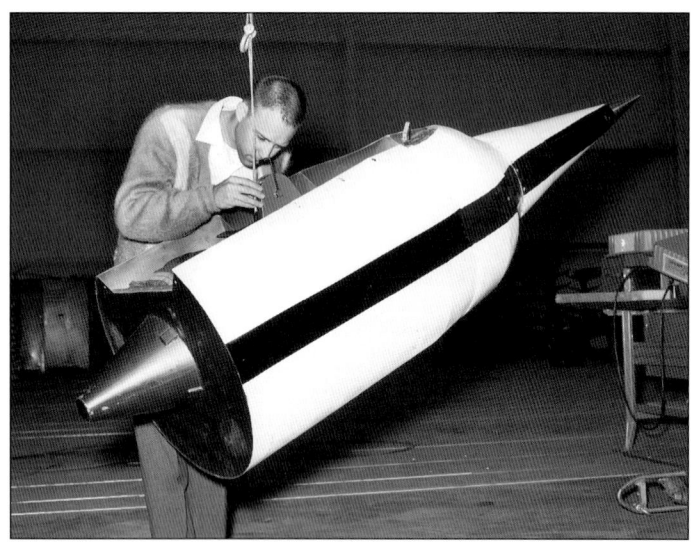

In order to gain basic aerodynamic data and to investigate the effects of carrying a generic ramjet shape on the X-15A-2, the Flight Research Center had a series of "dummy" ramjets constructed. Two different nose configurations were flown, the first have a 20-degree cone and no probes, the second having a 40-degree cone and two probes from the extreme tip of the nose cone. The 20-degree nose cone was flown on Flights 2-51-92 and 2-52-96; the 40-degree cone was flown on Flight 2-53-97. The photo below left shows the ablative coating on the ramjet, and below right shows the ramjet after it was unexpectedly jettisoned from Flight 2-53-97. (NASA Dryden)

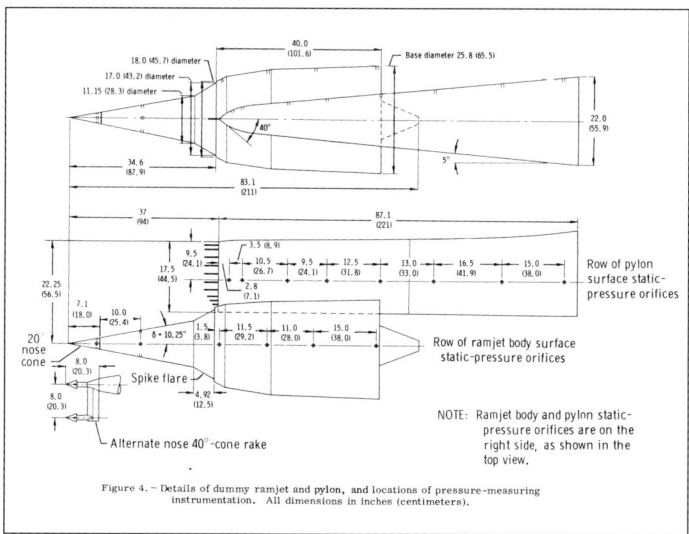

A good set of drawings detailing the dummy ramjet and the modified ventral stabilizer it was attached to. Note (at right) that the modified ventral has a straight leading edge (no sweep) along with ten pressure rakes. The relationship to the skid shows how much shorter the modified ventral was. The drawing above shows both the 20-degree cone and 40-degree cone variants of the dummy ramjet. (NASA)

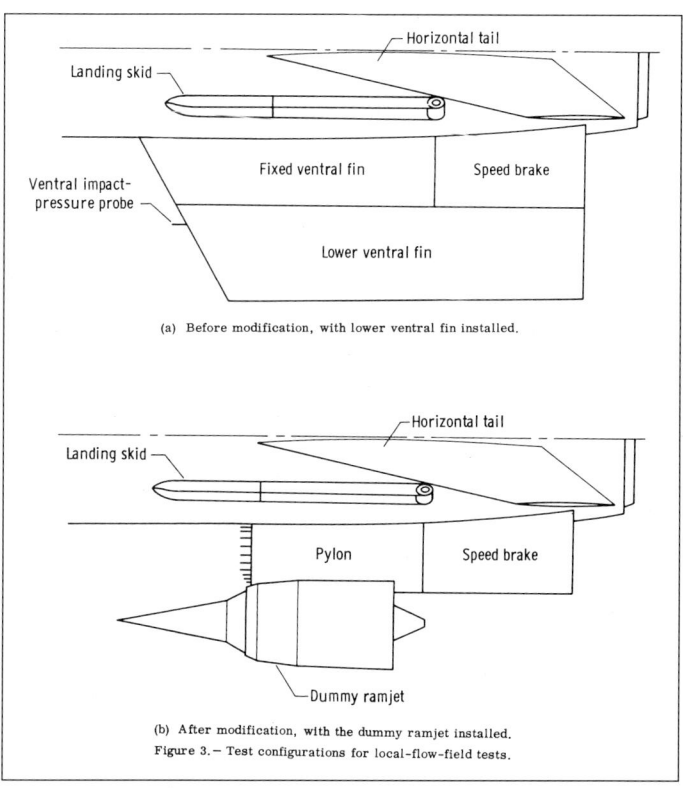

Pete Knight inspects the X-15A-2 on 4 August 1965 after the Martin MA-25S ablator has been applied. This marks the only full-scale use of an ablator on a manned aircraft, and the problems encountered during the program eventually led to it being dropped as a candidate for the Space Shuttle thermal protection system. (NASA Dryden)

Pete Knight and other concerned crewmembers gather around the tail of the X-15A-2 to inspect the damage to the ventral stabilizer after the Mach 6.7 flight on 3 October 1967. The damage had caused the dummy ramjet to separate long before Knight attempted to jettison it. (AFFTC History Office via the Terry Panopalis Collection)

X-15 Photo Scrapbook

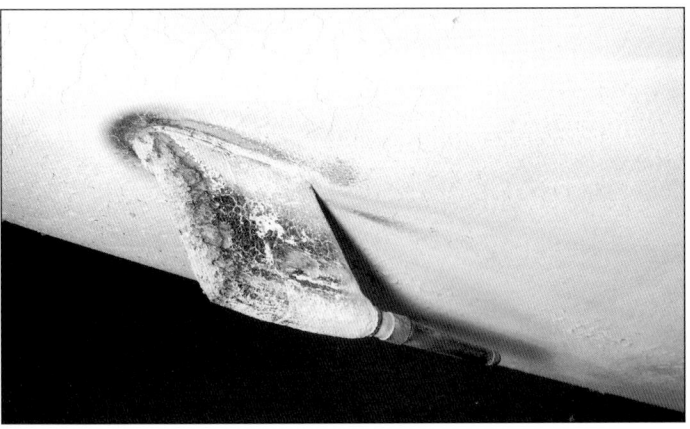

At top is a Martin technician spraying the MA-25S ablator onto the airplane. The photo above shows another technician sanding the ablator to the correct thickness. The full-scale ablator application on 6 June 1967 proved to be a lot of work. (NASA Dryden)

Above and Facing Page: *Various views from 3 October 1967 of the damage suffered by the X-15A-2 on its maximum speed flight. Surprisingly, with the exception of the ventral stabilizer, the damage was superficial and not much worse than expected.* (NASA Dryden)

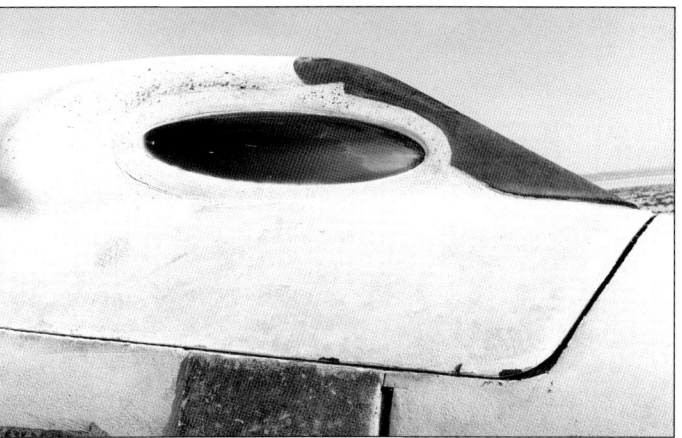
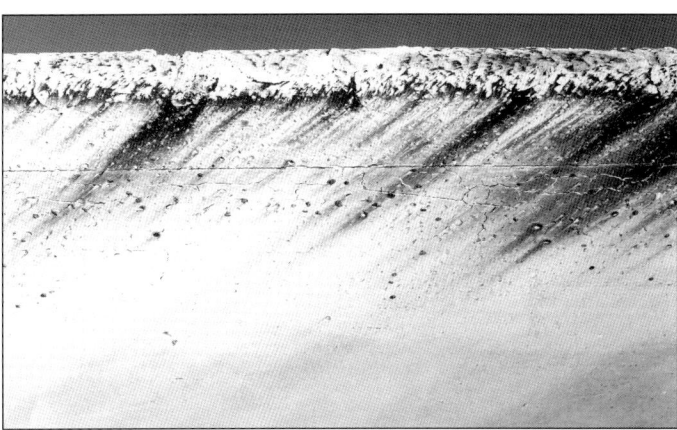

X-15 Photo Scrapbook

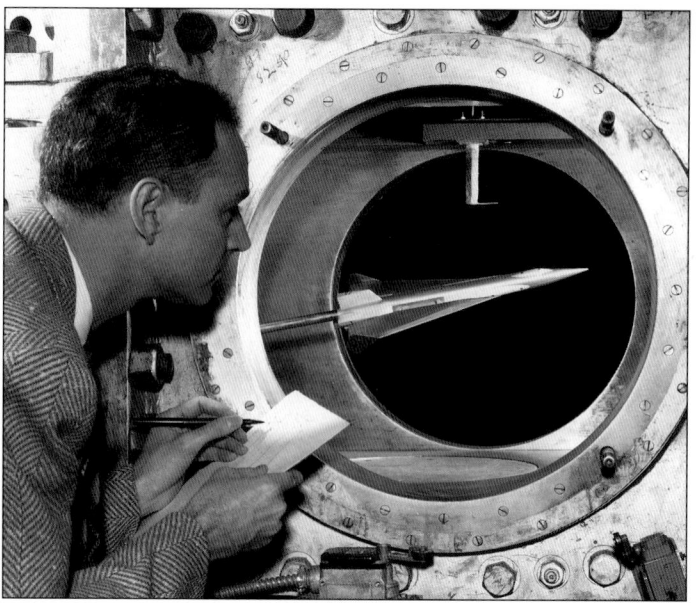

The X-15-1 returns from North American Aviation on 15 February 1967 under the watchful eye of a NASA security guard. The ammonia tank in the airplane had ruptured during Flight 1-69-116 on 6 October 1966, necessitating the airplane return to Inglewood for repair. (NASA Dryden)

Although it was never sanctioned as an official project, the delta-wing proposal underwent extensive development at both the Flight Research Center and at Langley. Here is a small-scale model of the final configuration in one of the Langley wind tunnels. (NASA Langley via the Terry Panopalis Collection)

Like many aircraft, the X-15s suffered from continual weight gain, and in an attempt to minimize landing loads a third skid was added under the ventral stabilizer on X-15-1 for its last 12 flights and on X-15-3 for its last 14 flights. (NASA Dryden)

Below: *The wreckage of X-15-3 after Flight 3-65-97 on 15 November 1967.* Above: *At left is the wing-tip pod where the electrical disturbance occurred that contributed to the accident. The smashed instrument panel is at right.* (NASA Dryden)

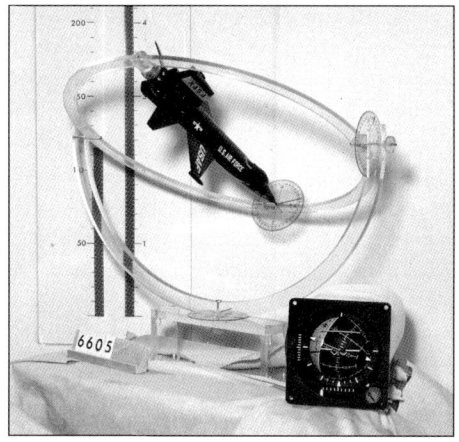

Either Side: *To reconstruct what happened during the last minutes of Flight 3-65-97, engineers constructed a very detailed X-15 model (note the tailcone box and wingtip pods) and a stand that allowed full freedom of movement. Based on telemetry data the model was positioned and photographed with a 16mm movie camera. When played in real time, this frame-by-frame reconstruction provided an excellent visual representation of what Mike Adams had gone through. These photos were taken on 22 December 1967; the one at right shows the entire setup, while the one at left shows the model.* (NASA Dryden)

X-15 Photo Scrapbook

Above and Facing Page: *By March 1968, only the X-15-1 remained on flight status; the X-15A-2 had been grounded after its maximum speed flight, and X-15-3 had been lost the previous November, killing Mike Adams. This is X-15-1 after the first flight (1-74-130) following Adam's accident, on 1 March 1968. Note the tail-cone box, but the lack of wing-tip pods. Several different tail-cone boxes had been built during the program, ranging from uncomplicated structures that simply covered the instruments within them, to elaborate devices equipped with stabilized platforms to allow cameras and other instruments to precisely track the horizon. The particular box held the MIT-Apollo Horizon Photometer Phase II experiment (#17) with a spectral photometer, a camera, and a star tracker mounted on a three-axis stabilized platform. The star tracker was designed to acquire Polaris and the gimbaled system could control horizon scan rates to obtain the most useful data independent of aircraft maneuvering. All of the instruments were covered by a door during the X-15 boost phase; the door was opened above 100,000 feet. Note the tag on the XLR99 engine – it says, in typical military parlance, "Engine, Rocket, Liquid Propellant."* (NASA Dryden)

At the end of the program, the X-15A-2 was used for a series of heat simulation tests during November 1968. This shows the setup for the horizontal stabilizer (left) and the right wing (right). (NASA Dryden)

1/15-scale X-15A-2 (left) and delta-wing (right) wind tunnel models displayed at the Air Force Flight Test Center Museum at Edwards AFB. The delta wing model is representative of the final proposed configuration. Photographed on 5 August 2002. (Tony Landis)

Bill Dana in his A/P22S-2 full-pressure suit rides the U.S. Savings Bonds float in the Rose Parade on 1 January 1968. (NASA Dryden)

As initially delivered, the interior surface of the Reaction Motors XLR99 thrust chamber was coated with a 0.005-inch-thick Nichrome undercoat covered with 0.010-inch of Rokide Z zirconia. This coating began to flake off during the flight program. A new coating was developed that consisted of 30 mils of molybdenum primer in the throat and 10 mils elsewhere followed by 6 mils of a graduated nichrome-zirconia coating, and then 6 mils of a zirconia topcoat. A special jig (right) was installed in the rocket engine shop at Edwards to allow the chambers to be recoated in the field. The rotary jig turned the chamber in an arc while the coating was sprayed on under high pressure and heat. (U.S. Air Force via the Terry Panopalis Collection)

Balls Three carries the X-15-3 and Bob White aloft on 17 July 1962 for Flight 3-7-14. (NASA via the Terry Panopalis Collection)

Note the H-21 rescue helicopter in the background during a landing of the X-15-2 on 1 June 1962. (Terry Panopalis Collection)

Bob Rushworth brings the X-15-3 home after his astronaut qualification flight on 27 June 1963. (NASA via the Terry Panopalis Collection)

When the Saturn insulation was carried (left, on 28 March 1968), each wingtip pod had a camera installed in the back portion aimed at the insulation. The cameras could be pointed at the upper or lower speed brakes. The general idea was to get the airplane into the desired region of dynamic pressure and temperature, then deploy the speed brakes to determine the effects on the insulation and its adhesives. Exactly what was learned through these experiments could not be ascertained. (NASA Dryden)

X-15 Photo Scrapbook

Mike Adams' family on 8 May 1992 after Adams was inducted onto the Astronauts Memorial at the Kennedy Space Center in Florida. From Left: Georgia Adams Baker (Mother), Roland Borden, Liese Adams Borden (Daughter, with baby Lane Aliese), Freida Adams Brian (Wife), Brent Adams (Son), George Adams (Brother), Mike Adams (Son). The invitation is at center and right.
(left: courtesy of Freida Adams Brian; right: Dennis R. Jenkins Collection; above: John Gourley)

Interesting only because the original large serial number on the upper rudder may be seen through the blistered paint after a high speed flight. Note the small serial on the fixed portion of the stabilizer. The photo is dated 13 May 1960. (NASA Dryden)

The Aeronautical Order awarding Pete Knight his Astronaut Wings for Flight 3-64-95 on 17 October 1967. According to the U.S. military, flights above 50 miles (264,000 feet) qualified a pilot as an astronaut. Mike Adams (1 flight), Joe Engle (3 flights), Pete Knight (1 flight), Bob Rushworth (1 flight), and Bob White (1 flight) therefore received Astronaut ratings from the Air Force for their X-15 experiences. On the other hand, NASA abided by the international (FAI) standard that says space begins at 100 kilometers (62 miles, 328,099 feet). Under these rules, Joe Walker should have qualified for a NASA Astronaut rating (for two flights), but NASA has steadfastly refused to recognize the achievement. Two other NASA pilots, Bill Dana and Jack McKay, flew above 50 miles, but did not exceed 62 miles. NASA is perfectly willing to call people astronauts that have never flown in space as long as they are assigned to the Shuttle program, but ignores a pilot who actually flew into space twice. (Courtesy of Pete Knight)

The X-15-1 hangs from the ceiling in the main gallery of the National Air and Space Museum in Washington D.C. The airplane is mostly as it appeared on its last flight, complete with wing-tip pods, but lacking the tail-cone box. (Tony Landis)

By 7 May 1969 the X-15-1 had been disassembled for shipment to the Smithsonian, although exactly how it was transported remains a mystery. Most probably the Air Force provided a C-141 or C-5 to airlift the airplane to Dulles or Andrews AFB, then it was trucked overland. A similar mystery surrounds how it got back to the desert for display in 1972 (see photo at right). (NASA Dryden)

The X-15-1 was loaned (from the Smithsonian) to NASA and put on display at the Flight Research Center's 25th Anniversary celebration on 8 September 1972. The airplane languished in a storage yard at the center for several years before being returned to the Smithsonian in time for the opening of the museum on the Mall. (NASA Dryden)

For several years after its arrival at the Air Force Museum the X-15A-2 was displayed in its natural black metal finish without markings. Later, correct markings were applied, and a set of external tanks added. In 2003 the airplane was moved out of the main museum building into the Annex (right) across the field where it is difficult to get to. (Above: Jay Miller; two at right: Tom Tullis)

The X-15 replica used in the movie "X-15" was subsequently modified to the A-2 configuration and is now on display at the Pima Air Museum in Tucson. The replica is indoors, and this photo is from September 2001. Although no conclusive evidence could be found, it is likely that this was the original X-15 engineering mockup used by North American Aviation. (David Allison)

Milt Thompson believed the Flight Research Center needed to have an X-15 replica to remind the employees of their rich heritage in exploration. This replica of the X-15-3 was constructed in 1992-93 for DFRC and is now mounted on a pole in front of the center. (above: NASA Dryden; right: Dennis R. Jenkins)

Two views of the NB-52A pylon under construction. The pylon was a decidedly simple structure, although it was very robust to handle the weight of the X-15. Early in the flight program it was modified to house a liquid nitrogen cooling system for the stable platform. (The Boeing Company Archives)

Balls Three and the X-15-1 around the time of the first captive flight. Note that the NB-52A does not have its launch operator's astrodome installed yet and that there is no nose art on the forward fuselage. The early international orange markings were very colorful. (The Boeing Company Archives)

The color has shifted, but this is a great period shot showing two engineers in front of Balls Three. Note Scott Crossfield in his David Clark Company MC-2 full-pressure suit in the backgrounds. The NB-52A has its astrodome, but still no nose art. (Gerald H. Balzer Collection)

The brand new X-15-1 poses for its official portrait around the time of the roll-out ceremony in October 1958. Note the NACA air data boom on the nose and the bug-eye camera fairings behind the cockpit. (The Boeing Company Archives)

One of the few color images from the centrifuge training at NADC Johnsville. All of the original cadre of X-15 pilots and flight planners spent a great deal of time at Johnsville, with Scott Crossfield (pictured) leading the charge for North American. The experience proved beneficial for all involved. (U.S. Navy)

The HL-10 (left) and X-15A-2 on 1 May 1966. Both aircraft had an influence on the development of the Space Shuttle. (NASA Dryden)

The X-15-3 ignites its XLR99. The white "smoke" is condensation from the very cold propellants. (NASA via the Terry Panopalis Collection)

The X-15A-2 at Mud Lake after the first attempt (2-45-81) to fly using the external tanks on 1 July 1966. (AFFTC History Office)

Neil Armstrong and Milt Thompson on 26 September 1962. Milt is congratulating Neil on his selection into the second group of astronauts, something that would eventually lead to a Gemini flight and a trip to the Moon for Armstrong. (NASA Dryden)

Scott Crossfield, Iven Kincheloe, and Joe Walker with the X-15 mockup, probably in early 1958. (The Boeing Company Archives)

Scott Crossfield (left) hands over the "keys" to Neil Armstrong while Bob White looks on. This was the ceremonial turnover of the program from North American to the government. (AFFTC History Office via the Terry Panopalis Collection)

Not directly an X-15, but related and interesting nonetheless. At right is Joe Walker holding a certificate (shown above) he was given after a 90,000-foot flight in the JF-104. The photo is dated 21 June 1960. Walker would not exceed this altitude in the X-15 until his sixth flight (2-14-28) on 30 March 1961. (NASA Dryden; certificate courtesy of Mrs. Grace Wiesmann)

X-15 Photo Scrapbook

Wing-tip pod on X-15-1 on 15 October 1964. The Atmospheric Density Measurements experiment (#12) is shown in its deployed position from the nose of the pod. The back of the pod held the Micrometeorite Collection experiment (#13). (NASA Dryden)

The XB-70A-1 and X-15A-2 pose at Edwards in 1967. The B-70, also a North American design, weighed almost ten times as much as the X-15, but could still travel over 2,000 mph (Mach 3+). (NASA Dryden)

Martin MA-25-S ablator on a nose panel on X-15-1 on 10 December 1964. These nose panels were frequently used to test materials since they were removable and could be easily instrumented. Also note the details of the BCS thrusters. (NASA Dryden)

The X-15A-2 at its roll out. (Boeing via the Terry Panopalis Collection)

An XLR99-equipped X-15 under the wing of Balls Three during November 1961. (Dean Conger)

Joe Engle with his family after Flight 3-44-67 on 29 June 1965. This was Engle's Astronaut qualification flight. (NASA Dryden)

One of the X-15s sitting at the Rocket Engine Test Facility at Edwards. (AFFTC History Office)

The X-15-3 on 2 April 1963. At this point the airplane had not flown since 18 January, mainly because of rain in the area that had made the lakebeds unusable, a frequent problem during the winter. Joe Walker would make the next flight (3-15-25) on 18 April. (NASA Dryden)

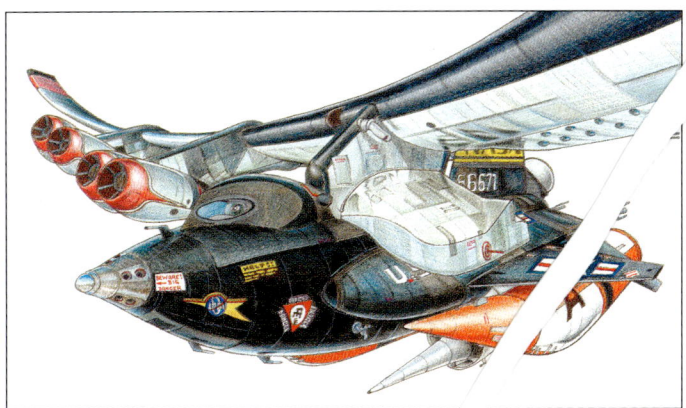

Ronald Stephano in Luxembourg draws caricatures of his favorite aircraft. Here is the X-15A-2. (©2003 Ronald Stephano)

X-15 Photo Scrapbook

Wind direction on the lakebed was determined by watching smoke from flares set up along side the runway. This is Joe Engle returning from Flight 3-35-57 on 28 September 1964 with Pete Knight in the F-104 as Chase-4. (AFFTC History Office)

Milt Thompson leaning against the mock-up of X-15-3 that was built largely at his urging and was installed on a pole outside the Dryden Flight Research Center. (NASA Dryden)

Right: *The cockpit of X-15-3 on 1 July 1966, after the Lear instrument panel was installed. The Ames Boost Guidance display is in the lower center portion of the panel. Note the advanced vertical tape-style instruments, but the stopwatch is still centerstage.* (NASA Dryden)

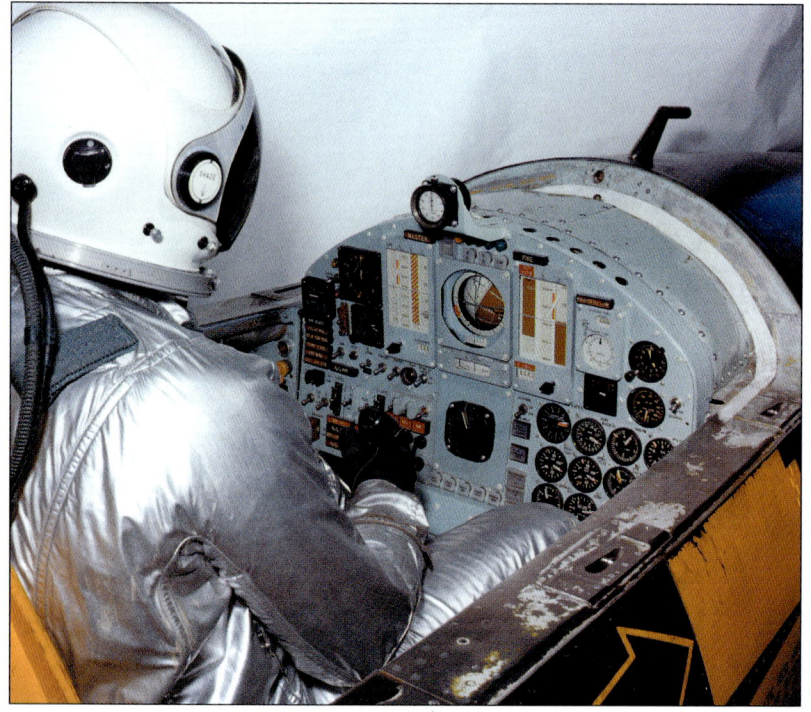

Bob Rushworth would be the first person to test the X-15 without the ventral rudder during Flight 1-23-39 on 4 October 1961. The concept seemed to work well, but the program waited until the aerodynamicists finished running their wind tunnel tests before committing to a second flight without the ventral. Eventually, the program would make 73 flights with the ventral and 126 flights without it. *(NASA Dryden)*

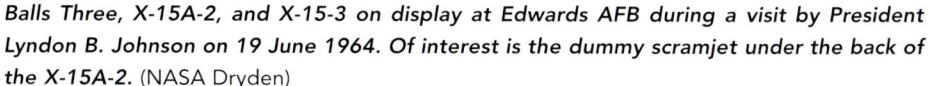

Balls Three, X-15A-2, and X-15-3 on display at Edwards AFB during a visit by President Lyndon B. Johnson on 19 June 1964. Of interest is the dummy scramjet under the back of the X-15A-2. *(NASA Dryden)*

X-15 Photo Scrapbook

Dean Conger was a photographer with National Geographic for 30 years, and photographed the X-15 and its pilots on several occasions. The Air Force would not allow Mr. Conger to fly in the NB-52, so he remembers "the shots from the B-52 of the X-15 under the wing and dropping were with a remote camera I was able to have placed in the small bubble on the right side of the B-52. I had to have a special bracket made so a motor drive camera could be mounted in that bubble. It took a bit of figuring to take into account that the B-52's wing curves up some 20 feet in flight. I framed the pix and T/Sgt Robert L. Wynn triggered the motor drive, while I was in a chase plane." The other images of the airplane and pilots, and those of the Collier Trophy presentation at the White House, were shot by Mr. Conger without the intermediary. The images on these two pages were provided through the courtesy of Mr. Conger.

A beautiful shot of Balls Three with an F-100 (Chase-1) tucked into close formation. Note the use of international orange on both aircraft. (Dean Conger)

Since most X-15 flights were scheduled for the early morning, preparations were frequently made under the glare of spotlights. Here is X-15-1 and Balls Three in 1959. (Dean Conger)

Joe Walker tucked into an X-15 in October 1961. Note the large vent pipe coming out the side of the NB-52 pylon. (Dean Conger)

The Collier Trophy presentation at the White House. Crossfield, Walker, and White are just to the right of Kennedy. (Dean Conger)

Joe Walker in his David Clark full-pressure suit prior to a flight (probably 1-24-60) in late 1961. (Dean Conger)

An F-104 chase plane zooms into position just before the X-15 is launched from Balls Three. (Dean Conger & T/Sgt. Robert L. Wynn)

Joe Walker, Scott Crossfield, and Bob White behind the Collier Trophy, still inside the White House. (Dean Conger)

Major Robert M. White during the ceremony when he received his Astronaut Wings in November 1961. (Dean Conger)

A very relaxed-looking Joe Walker. The remnants of thermal paint may be seen on the nose of the X-15. (Dean Conger)

X-15 Photo Scrapbook

Views of the X-15A-2 on various flights showing the paint scheme on the external tanks. Top row from left: Flight 2-45-81 on 1 July 1966, 2-50-89 on 18 November 1966, 2-C-93 on 7 August 1967 (a scheduled captive flight to check out the tanks). This row from left: Flight 2-52-96 on 21 August 1967 (no tanks carried) and 2-53-97 on 3 October 1967. All the tanks were recovered, although not always in a reusable condition. (NASA Dryden and AFFTC History Office)

Balls Three and the X-15A-2 complete their preflight checks for Flight 2-43-75 on 3 November 1965. The main Edwards control tower is to the left and the hangar built for the B-70 program is in the background. (NASA Dryden)

Views immediately after the X-15A-2 landed on its maximum speed flight. The airplane suffered significant ablator damage in some areas, especially the ventral stabilizer. Pete Knight is in the photo at right, still in his A/P-22S-2 full-pressure suit. The photo below shows the ground crew venting the propellant tanks. (SSgt. Robert Hoffman)

X-15 Photo Scrapbook

Yes, that is snow at Edwards AFB. The gang gathers after the last attempt at the 200th flight is called off on 20 December 1968, marking the end of the X-15 program. We know the the same photo was used in Hypersonic, but we thought you might enjoy seeing it in color.

Standing, left to right: Donald L. Hall, Cyril G. Brennan, Gilbert W. Kincaid, Robert L. Schuck, Gaston A. Moore, Vincent N. Capasso, Jr., Paschal "Herm" Dorr (8mm camera), Allen F. Dustin, Harvey B. Price, Nicholas Kantartzis, Richard H. Simon, James B. Craft, George E. Perusich, Edward T. Ryan, Walter P. Redman, unknown, and Ira O. Cupp. On the stairs, left to right: Richard L. Blair and Marshall E. McCracken. (Dryden History Office Collection)